Analytical Techniques for the Elucidation of Protein Function

Analytical Techniques for the Elucidation of Protein Function

Edited by

Isao Suetake
Nakamura Gakuen University, Japan

Rohit K. Sharma
Panjab University, India

Hironobu Hojo
Institute for Protein Research, Osaka University, Japan

WILEY

This edition first published 2023

Registered Office(s)
John Wiley & Sons, Inc., 111 River Street, Hoboken, NJ 07030, USA
John Wiley & Sons Ltd, The Atrium, Southern Gate, Chichester, West Sussex, PO19 8SQ, UK

For details of our global editorial offices, customer services, and more information about Wiley products visit us at www.wiley.com.

Wiley also publishes its books in a variety of electronic formats and by print-on-demand. Some content that appears in standard print versions of this book may not be available in other formats.

Library of Congress Cataloging-in-Publication Data
Names: Suetake, Isao, editor. | Sharma, Rohit K., editor. | Hojo, Hironobu, editor.
Title: Analytical techniques for the elucidation of protein function / edited by Isao Suetake, Rohit K. Sharma, Hironobu Hojo.
Description: Hoboken, NJ : John Wiley & Sons, 2023. | Includes bibliographical references and index.
Identifiers: LCCN 2022038118 (print) | LCCN 2022038119 (ebook) | ISBN 9781119886327 (hardback) | ISBN 9781119886334 (pdf) | ISBN 9781119886341 (epub) | ISBN 9781119886358 (ebook)
Subjects: LCSH: Proteins. | Electron paramagnetic resonance spectroscopy. | Nuclear magnetic resonance spectroscopy.
Classification: LCC QP551 .A485 2023 (print) | LCC QP551 (ebook) | DDC 572/.6--dc23/eng/20221005
LC record available at https://lccn.loc.gov/2022038118
LC ebook record available at https://lccn.loc.gov/2022038119

Cover Image: Courtesy of Tomohiro Hojo
Cover Design: Wiley

Set in 9.5/12.5pt STIXTwoText by Integra Software Services Pvt. Ltd, Pondicherry, India

Printed and bound by CPI Group (UK) Ltd, Croydon CR0 4YY

C9781119886327_231222

Contents

Preface

Proteins are fundamental biomolecules in living organisms and play various essential roles. They are composed of twenty amino acid connected by amide bonds and form secondary structures, such as α-helix and β-sheet, and ultimately, tertiary structures (also quaternary structures in the case of protein complexes). As the structure of proteins is very important for their function, various methods have been used to analyze them. The first structure clarified was myoglobin, which was reported in 1958 using X-ray crystallography. Nuclear magnetic resonance spectroscopy has also been highly advanced and widely used for protein structural analysis. Recently, cryoelectron microscopy has also become a powerful tool for this purpose. The structural data is deposited on the Protein Data Bank, and the number of solved structures is rapidly increasing. As the data is accessible from all over the world, it supports the progress of research on protein as well as the development of therapeutics.

Recently, the region in a protein that does not assume a particular three-dimensional structure, namely, the intrinsically disordered region (IDR), has been attracting much attention due to its importance for protein functions. IDR tends to form a liquid-liquid separated structure, which provides a site for various biological processes. IDR often receives various post-translational modifications, which are important to modify the function of proteins. In addition, the post-translational modifications often occur heterogeneously. To analyze the structure and function of these regions, the use of the above mentioned methods, such as X-ray crystallography and NMR, are not sufficient, as the regions are highly heterogeneous, mobile, and do not have firm tertiary structures. In addition, the analysis to identify the site and the kind of post-translational modification is required to elucidate how these modifications play roles.

In this book, we want to provide a brief introduction to less popular yet promising techniques to undertake functional and structural analyses of proteins, especially the intrinsically disordered region and the post-translational modifications.

We anticipate that many of the readers are not familiar with the described techniques nor their theoretical background. Therefore, the authors describe each technique starting from a simple introduction and including a theoretical background, followed by the application of the method to the analysis of protein structure and function.

Chapter 1 describes the electron paramagnetic resonance spectroscopy technique. The method can be applied to the analysis of long-range interaction and fast motion of the proteins by specific spin labeling. Neutron scattering, described in Chapter 2, is an efficient method to analyze the dynamics of proteins. The method is particularly effective to analyze the dynamics of membrane proteins in lipid bilayers. Chapter 3 deals with Raman spectroscopy, which can clarify biological processes, such as protein-protein interaction and folding, by the analysis of scattering at the specific wave number indicative of a specific bond. This can be done with a small amount of sample without labeling. In Chapter 4, the author describes that the structure and aggregate formation of proteins can be globally analyzed by impedance measurement of a protein solution, which can be done without introducing any probes. Mass spectrometry imaging, described in Chapter 5, can analyze the distribution of various compounds in tissue samples with extreme high sensitivity using mass spectrometry. This method can be applied for the analysis of a wide range of compounds, as it identifies the target compounds by their molecular weight. Chapter 6 describes the single-molecule monitoring by quantum dots, which achieves the analysis of protein molecules one by one and is clearly advantageous when proteins with heterogeneous post-translational modifications are analyzed. Chapter 7 deals with solid-state NMR. The method can analyze proteins even in an aggregated state and micelles, which is very effective for the analysis of amyloid formation and membrane proteins in lipid bilayers. In Chapter 8, the novel method for the introduction of materials into cells as well as protein crystallization using bubble knife is described. These techniques are of great use for the analysis of proteins, which are difficult to be treated by conventional analytical methods.

We would like to express our sincere thanks to all the authors who contributed to this book. We also appreciate the efforts of reviewers who helped revise the text. Our thanks also go to Ms. Jenny Cossham and Ms. Elke Morice-Atkinson of John Wiley & Sons for their continuous help to realize the publication of this book.

We hope that the book will be a good introduction to the described techniques and contribute to increase their popularity for protein analysis, especially through research on the IDR. Also, we hope that the book will contribute to the further advance of protein science in the future.

Isao Suetake
Rohit K. Sharma
Hironobu Hojo

Editor's Biographies

Isao Suetake, professor at the Graduate School of Nutritional Sciences, Nakamura Gakuen University, Japan.

Isao Suetake is a professor of nutrition at Nakamura Gakuen University. He graduated from the Faculty of Science, Osaka University, in 1990 with a bachelor's degree and got his PhD in 1996 from the university. He specializes in biochemistry and molecular epigenetics and has researched the properties of enzymes responsible for DNA methylation and molecular mechanisms recognizing histone modifications. For more than ten years, he has collaborated with Prof. Hojo, who is one of editors of this book, to combine chemistry and bioscience for elucidating protein modifications.

Rohit K. Sharma, assistant professor at the Department of Chemistry, Panjab University, India.

Rohit K. Sharma obtained his PhD in medicinal chemistry, National Institute of Pharmaceutical Education and Research (NIPER) in 2009. His thesis was on the design and synthesis of antimicrobial peptides. In 2007, he underwent a research sojourn in the Helmholtz Centre for Infection Research (HZI), Germany. In 2009, he moved to Nanyang Technological University, Singapore, as a postdoctoral research fellow. He moved to his present position in Panjab University in 2011, where his research interests include understanding nano-peptide conjugates and self-assembling organic networks with applications in drug delivery and bio-sensing.

Hironobu Hojo, professor at the Institute for Protein Research, Osaka University, Japan.

Hironobu Hojo obtained his PhD in organic chemistry from Osaka University in 1994. His thesis was on the development of the method for chemical protein synthesis, the thioester method. In 1998, he moved to Tokai University and started to extend the thioester method for glycoprotein synthesis. He moved to his present position, professor at the Institute for Protein Research, Osaka University, in 2013 and is developing a chemical approach toward the understanding of the function of post-translationally modified proteins.

List of Contributors

Toshimichi Fujiwara
Osaka University
Suita, Osaka
Japan

Hironobu Hojo
Osaka University
Suita, Osaka
Japan

Gurpreet K. Soni
Panjab University
Chandigarh
India

Alisha Lalhall
Panjab University
Chandigarh
India

Saima Malik
Panjab University
Chandigarh
India

Tatsuhito Matsuo
Université Grenoble Alpes
Grenoble
France

Hiroyuki Mino
Nagoya University
Nagoya
Japan

Judith Peters
Université Grenoble Alpes
Grenoble
France

Deepika Sharma
Panjab University
Chandigarh
India

Isao Suetake
Nakamura Gakuen University
Fukuoka
Japan

Rohit K. Sharma
Panjab University
Chandigarh
India

Shuichi Shimma
Osaka University
Suita, Osaka
Japan

Maitrayee U. Trivedi
Panjab University
Chandigarh
India

Nishima Wangoo
Panjab University
Chandigarh
India

Yusuke Tsutsui
Kyoto University
Kyoto
Japan

Yoko Yamanishi
Kyushu University
Fukuoka
Japan

1

EPR Spectroscopy

Analytical Techniques for the Elucidation of Protein Function, First Edition. Edited by Isao Suetake, Rohit K. Sharma, and Hironobu Hojo.

1.1

Outline of EPR Spectroscopy

Hiroyuki Mino

Division of Material Science, Graduate School of Science, Nagoya University, Furo-cho, Chikusa-ku, Nagoya, Aichi, Japan

1.1.1 Overview

Our most familiar interaction in this world is electromagnetism. Almost all phenomena in material physics arise fromelectromagnetic interactions between light and matter. Our world is a stage for the electron as the main actor. Biosystems are also standing on the stage.

Electromagnetism is mediated by electromagnetic waves, called photons, which are the particles of the force field (the bosons, or force carriers). The electromagnetic waves are classified by frequency: γ-rays, X-rays, ultraviolet light, visible light, infrared light, microwave, and radio wave, in order of decreasing energy. Optical absorption spectroscopy using visible light is the most basic spectroscopic tool. Electron paramagnetic resonance (EPR) is also a spectroscopic tool [1–3]. Optical absorption detects the transition of the electron dipole moment, while EPR detects the transition of spin angular momentum. Spin (S) is classically modeled by the behavior of a small bar magnet. In quantum mechanics the bar magnet is allowed to be only in two states, parallel or antiparallel to the external magnetic field. In the simple case of $\boldsymbol{S} = 1/2$, only the transition between two states, the up-spin state, $|\alpha\rangle$ ($m_I = -1/2$), and down-spin state, $|\beta\rangle$ ($m_I = 1/2$), are allowed. Under an external magnetic field, energy levels are separated, called Zeeman interaction (Figure 1.1.1). The transition between two states is performed by microwave irradiation. Since the two transitions, $|\alpha\rangle$ to $|\beta\rangle$ and $|\beta\rangle$ to $|\alpha\rangle$, have the same probabilities, the periodic transitions between the $|\alpha\rangle$ and $|\beta\rangle$ states continue during microwave irradiation and is called

Analytical Techniques for the Elucidation of Protein Function, First Edition. Edited by Isao Suetake, Rohit K. Sharma, and Hironobu Hojo.

"Rabi oscillation" (Figure 1.1.2). This oscillation is not unique for magnetic resonance but general for all spectroscopies. If we can detect very fast measurements, the oscillation should be observed in optical light spectroscopy.

The energy for the spin transition might be smaller than that for the other electronic transitions. However, the electron spin is coupled with the orbital symmetry in the molecules. Therefore, the spin states determined by the EPR method are an indispensable tool to determine the quantum properties of the molecular orbits.

EPR and nuclear magnetic resonance (NMR) are categorized as magnetic resonance techniques. The differences between EPR and NMR are mainly ascribed to the spin size. Considering spin interaction between the spins, the effective interacting distance for a nuclear spin is around 10 Å, while that for electron spin is around 100 Å. Therefore, the EPR spectrum includes many magnetic interactions, such as the nuclei and electron spins within these distances (Figure 1.1.3). The difference in magnitude of the spins is also reflected in the

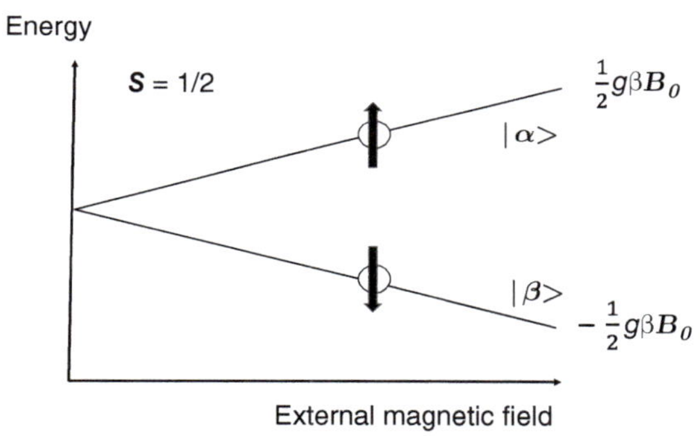

Figure 1.1.1 Energy levels for $S = 1/2$ spin in the magnetic field B_0.

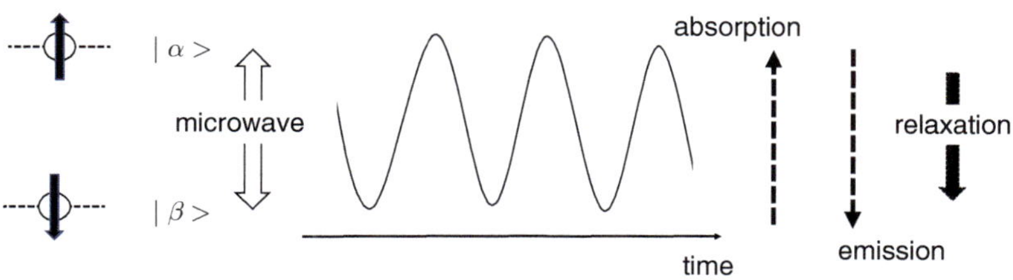

Figure 1.1.2 Illustration for the transitions in two levels |α> and |β>. Microwave irradiation gives oscillation between two levels (Rabi oscillation).

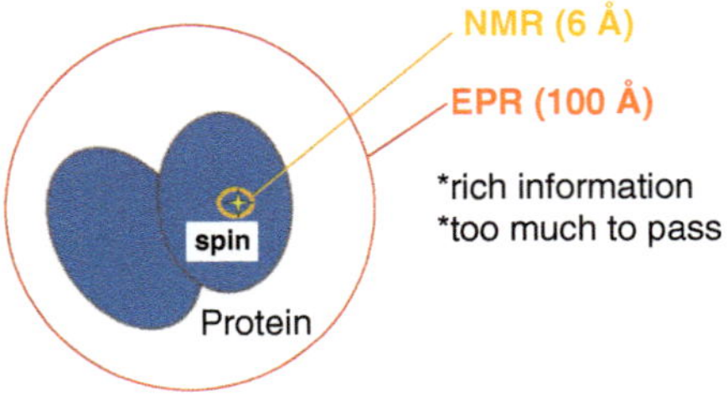

Figure 1.1.3 Scale for the interaction of the nucleus and electron spins relative to protein size.

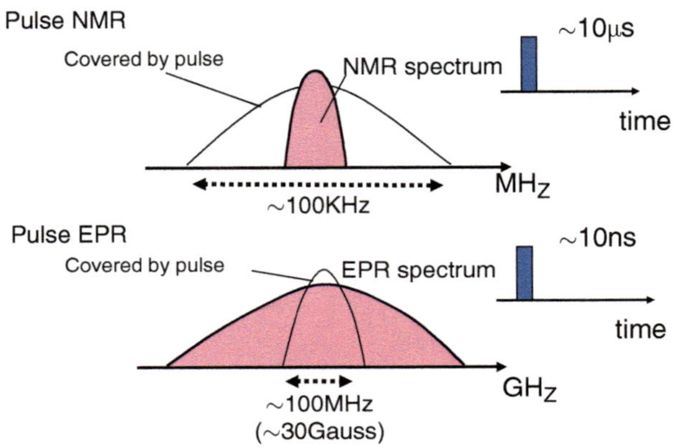

Figure 1.1.4 Coverage area for NMR pulse and EPR pulse relative to spectral width.

spectral linewidth. NMR is generally estimated as several kilohertz, while the EPR spectral width is estimated as 1 THz (0.3 nm). In the case of the visible light spectrum, it reaches about 300 THz (~100 nm). The differences directly reflect the pulsed technique. In pulse spectroscopy (Figure 1.1.4), the object is irradiated with a strong electromagnetic wave in a short period. The detectable spectral linewidth depends on the pulse width and intensity. For example, the pulse with a length of 10 μs covers a 100 kHz width in the spectrum. Therefore, the short pulse covers the whole NMR spectrum, while only a small region in the spectrum is detected in EPR and visible optical spectroscopy. In the case of EPR, the limitation of a strong microwave pulse is around 20 ns, indicating a linewidth of 50 MHz, about 20 G, in the spectrum. Therefore, EPR may be an intermediate methodology between NMR and visible light spectroscopy (Figure 1.1.5). The pulsed EPR technique has two options, whether to detect a wide range in the spectrum or conversely to detect frequency selective beaching in a spectrum (a hole), denoted as "Hole burning."

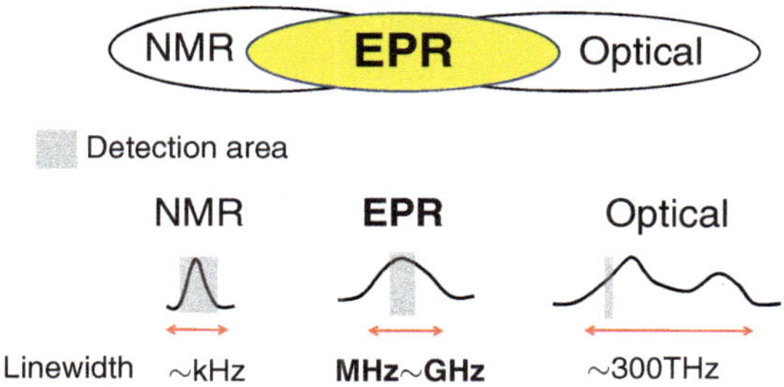

Figure 1.1.5 Comparison of EPR spectrum linewidth with NMR and optical spectral linewidth. Gray marker shows the detectable area for a single-pulse excitation.

The application to protein structure studies contributes to the advancement and popularity of the NMR technique. Protein studies are also an important application for EPR. The slow spin relaxation in the protein is suitable for EPR measurements, where the protein works as a kind of solvent virtually. As the electron spin interaction covers proteins, a lot of information in the spectrum disturbs the extraction of the necessary information. Therefore, many EPR techniques have been proposed.

In the external magnetic field $\boldsymbol{B}_0$ along the z-axis, the energy for a bar magnet with a magnetic moment $\boldsymbol{\mu}$ is described as:

$$\mathcal{H} = \mu B_0 \cos\theta, \tag{1.1.1}$$

where θ is the angle between the $\boldsymbol{B}_0$ and $\boldsymbol{\mu}$ vectors. The angle θ is arbitrarily selected in classical mechanics. But in quantum mechanics, only two states, parallel or antiparallel to the external magnetic field, are allowed. Using quantum mechanics ($\mu = g\beta\boldsymbol{S}$), the formula is rewritten as:

$$\mathcal{H} = g_e\beta_e B_0 S_z, \tag{1.1.2}$$

where g_e and β_e are the g-factor and Bohr magneton for the electron, respectively, and $S_z = 1/2$ or $-1/2$ (Figure 1.1.2). Considering the magnetic resonance in the two levels, the transition probabilities from the down spin to the up spin and from the up spin to the down spin are the same. As the number of lower levels (β) is larger than the number of upper levels (α) in the Boltzmann distribution, irradiated microwave energy is absorbed in the spin system. When the microwave energy is continuously supplied, the numbers of spin in both levels become equal, resulting in the loss of absorption. Therefore, the absorption signal is observed in the balance of suppling energy and spin relaxation. The evaluation of the spin relaxation gives the magnetic interaction between spins (Figure 1.1.2).

In the pulse EPR method, microwaves are irradiated only for a short time. Therefore, spin relaxation can be suppressed and evaluated. The electron-nucleus spin interaction can be directly observed by the ENDOR (electron nuclear double resonance) and ESEEM (electron spin echo envelope modulation) methods, and the electron-electron spin interaction is observed by the PELDOR (pulsed electron-electron double resonance) / DEER (double electron-electron resonance) [4] hole burning [5, 6], RIDME (relaxation-induced dipolar modulation enhancement) [7], etc.

When an external magnetic field is applied to the electron spin in the *z*-direction, the electron spin is quantized in the *z*-direction (Figure 1.1.6a). The electron spin inverts to the opposite direction by irradiating a microwave pulse with an appropriate length (Figure 1.1.6a). The inverted spin recovers with the thermal divergence to the outside. The recovery time is called spin-lattice relaxation time (T_1). T_1 is influenced by several different processes, such as the direct process, Raman process, and Orbach process, where the spin-spin interaction is also included. The T_1 measurement is one of the traditional distance measurement techniques.

Another relaxation time is the spin-spin relaxation time (T_2). The mechanism is classically expressed as spin relaxation on the *xy*-plane. As the spin is quantized along the *z*-axis direction in the external magnetic field, there is no energy dissipation in the *xy*-plane in the first order. The spin-spin relaxation is mainly described as the perturbation from the local magnetic field to the electron spin. When we consider the interaction between electron and nucleus spins, the local magnetic field from a nucleus spin is not always along to the *z*-axis. Therefore,

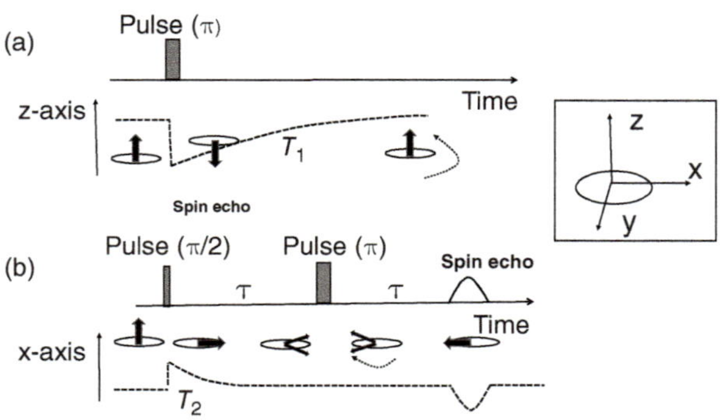

Figure 1.1.6 (a) Recovery of the magnetization to the *z*-direction in the rotation *xyz*-frame. After a single inversion pulse (π), the magnetization inverted to the −*z*-axis and recovered to the *z*-axis after T_1. (b) The two-pulse sequence with π/2 and π and magnetization irradiated along the *x*-axis. First, the magnetization is oriented from the *y*-axis to the *x*-axis after the π pulse. After the period of τ, the magnetization is dephasing on the *xy*-plane. Second, the magnetizations are refocused on the −*x*-axis after a period of τ.

fluctuations occur in the quantum transitions as phase relaxation. Inhomogeneity of the external magnetic field also contributes to T_2. Note that the single quantum relaxation, denoted as T_m (phase memory time), should be distinguished from the external factors. Figure 1.1.6b shows a basic pulsed technique for detecting spin echo. The *xyz*-frame is defined as the rotation frame, where the *xy*-plane is rotating around the *z*-axis in the Larmor frequency. Applying the first microwave pulse ($\pi/2$ pulse) with a microwave pulse from the *y*-axis, the spin magnetization is oriented along the *x*-axis, in the *xy*-plane. After time τ, the magnetizations are dephasing in the *xy*-plane. By applying a second microwave pulse (π pulse), the magnetizations are refocusing after time τ. The refocused magnetizations are called spin echo (Hahn echo). T_2 is obtained by measuring the τ-dependence. As the magnetization is quantized, spins are actually not oriented in the *xy*-plane, which simply consisted of the mixing of the up-spin and down-spin states. However, the classical mechanics model for spin dynamics is effective for quantum mechanics in many cases.

Spectral diffusion is another relaxation process, which is an exchange process in the spectrum consisting of inhomogeneous lines. Figure 1.1.7 shows a pulse sequence for the detection of spectral diffusion. The first soft pulse makes a sharp absorption hole in the spectrum. Afterwhile, the hole is dispersed. The dispersion of the hole can be monitored by a spin echo sequence using second and third pulses. As the process is due to the interaction with the local spin, we can evaluate the interaction. Notably, the spectral diffusion should be eliminated for T_1 measurements, and some useful techniques have been proposed for this purpose.

Magnetic dipole interaction is the basic interaction for magnetism. Although many kinds of complicated interactions are explained in the EPR books, most of the interactions are physically caused by the magnetic dipole interaction. As an

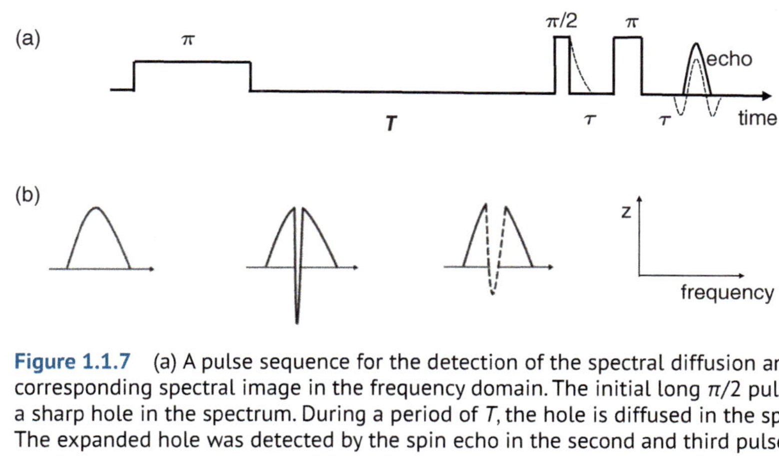

Figure 1.1.7 (a) A pulse sequence for the detection of the spectral diffusion and (b) corresponding spectral image in the frequency domain. The initial long $\pi/2$ pulse creates a sharp hole in the spectrum. During a period of T, the hole is diffused in the spectrum. The expanded hole was detected by the spin echo in the second and third pulses.

example, we consider the hyperfine interaction between an electron and a nucleus. The electron-nucleus magnetic dipole interaction is expressed by the following equation.

$$\mathcal{H}_{hyperfine} = -g_e\beta_e g_n\beta_n\left[\frac{(\boldsymbol{I}\cdot\boldsymbol{S})}{r^3} - \frac{3(\boldsymbol{I}\cdot\boldsymbol{r})(\boldsymbol{S}\cdot\boldsymbol{r})}{\boldsymbol{r}^5}\right] - \frac{8\pi}{3} g\beta g_n\beta_n\delta(\boldsymbol{r})\boldsymbol{I}\cdot\boldsymbol{S}, \quad (1.1.3)$$

where g_n and β_n are the g-factor and Bohr magneton for nuclei, respectively.

The first term is almost the same as the classical magnetic dipole interaction, as seen in a bar magnet. Although the second term is identified as the Fermi contact interaction, it is an exception for the calculation in the classical dipole interaction. In quantum mechanics, the electron is distributed in space, and therefore it is possible to locate it at the position of the nucleus, i.e. $\boldsymbol{r} = 0$, resulting in fraction divergence. Therefore, the second term is obtained from the first term after mathematical treatment for singularity.

The detection of hyperfine interaction is one of the main EPR subjects, which derive the molecular structure and orbitals. Techniques such as ENDOR and ESEEM are the specific tools to detect hyperfine interaction. In a narrow sense, the term "ESEEM" is used for the modulation effect of the hyperfine interaction. However, the term "ESEEM" does not specify the origin of the interaction. Note that the magnetic dipole interaction between electron spins is also denoted as ESEEM, which includes PELDOR/DEER and other techniques.

The hyperfine interaction is one of the most characterized features of the organic radical in CW EPR. Here is an example of a simple case, one unpaired electron ($\boldsymbol{S} = 1/2$) and one nucleus ($\boldsymbol{I} = 1/2$) coupled with only Fermi contact interaction. When the external field B_0 is directed to the z-axis, the energy is given by the following expression:

$$\mathcal{H} = g_e\beta_e B_0 S_z - g_n\beta_n B_0 I_z - aI_z S_z, \quad (1.1.4)$$

where the first term is the electron Zeeman interaction, already defined in Eq. (1.1.2), and the second term is the nuclear Zeeman interaction. The g_n and β_n are the g-factor and Bohr magneton for nuclei, respectively. The Fermi contact interaction a is obtained by the spatial integral of the second term of Eq. (1.1.5).

Figure 1.1.8 shows the energy diagram for the spin system. The energy levels branch into four states, $\alpha_e\alpha_n>, \alpha_e\beta_n>, \beta_e\alpha_n>, \beta_e\beta_n>$. The four energy levels are shifted by the hyperfine interaction a. The transitions for both spin states of S and I at the same time are forbidden. Therefore, two EPR transitions, $\nu_\alpha : \alpha_e\alpha_n > \Leftrightarrow \beta_e\alpha_n >, \nu_\beta : \alpha_e\beta_n > \Leftrightarrow \beta_e\beta_n >$ are allowed. The transitions are $g_e\beta_e B_0 + g_n\beta_n B_0 + 1/2a$ for ν_α and $g_e\beta_e B_0 + g_n\beta_n B_0 - 1/2a$ for ν_β. As a result, two EPR lines are separated with hyperfine interaction and centered with $g_e\beta_e B_0 + g_n\beta_n B_0$. In an $\boldsymbol{I} = 1$ system such as ^{14}N, three EPR lines separated with a

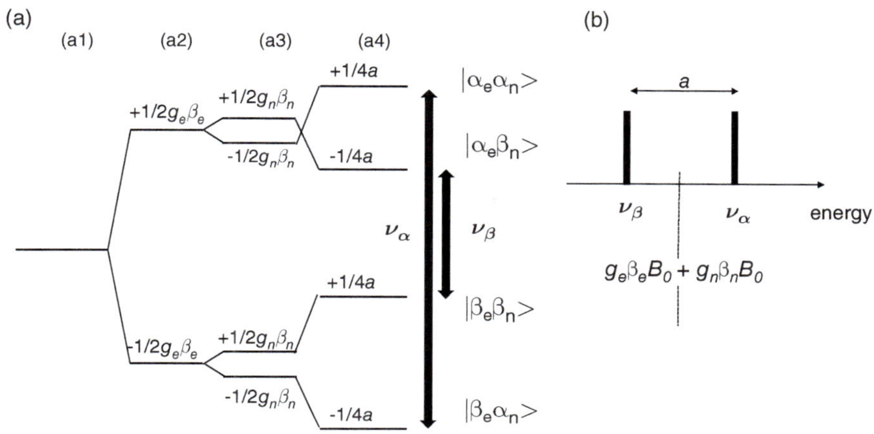

Figure 1.1.8 (a) Energy diagram for an $S = 1/2$ electron spin and an $I = 1$ nuclei spin with a Fermi interaction a. (a1) No magnetic field, Zeeman energy of (a2) electron, (a3) nuclei, and (a4) hyperfine interaction. (b) The spectral pattern for two allowed EPR transitions ν_α and ν_β on the energy axis.

are detected. If the spin system is composed of an unpaired electron and two nuclei ($I = 1/2$), four (2^2) EPR lines are detected. The number of EPR lines for N nuclei is 2^N. As the multi separations bury in the inhomogeneous lines, the EPR spectrum often becomes unresolved and featureless. ENDOR/ESEEM are good tools to extract the hyperfine interaction from the unresolved EPR spectrum because the number of lines becomes only 2N, where N is the number of nuclei in the ENDOR/ESEEM spectra.

The local magnetic field surrounding the electron spin derives modulation on the envelope. When the local magnetic field is due to nucleus spin coupling, the modulation pattern is described as the hyperfine interaction. The modulation frequency of the obtained nuclear modulation is determined by the data sampling interval, pulse width, and EPR linewidth, so the higher the frequency, the more difficult it is to observe. It is most effectively used for the observation of frequencies lower than 10 MHz (nitrogen nucleus, deuterium nucleus, etc.) in ESEEM.

Figure 1.1.9 shows the basic ESEEM spin sequence. Panels (a) and (b) are the two-pulse and three-pulse sequences. A modulation pattern is observed on the decay pattern of the spin echo in time dependence.

In the case of the two-pulse sequence, the envelope modulation is overlapping on the decay trace of the T_2 (T_m); therefore, the observable time is short. In the two-pulse ESEEM, the modulation pattern represents the multiplication of the two resonance frequencies, ν_α and ν_β, after Fourier transformation. In the case of the three-pulse sequence, the envelope modulation is overlapping on the decay

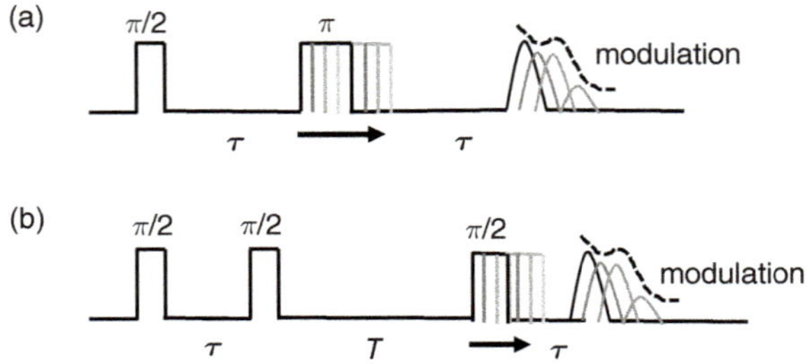

Figure 1.1.9 ESEEM pattern in (a) two-pulse sequence and (b) three-pulse sequence. By changing periods, the echo intensities are modulated.

trace of T_1, and therefore the observation time is relatively long. In the three-pulse ESEEM, the modulation pattern represents the summation of the ν_α and ν_β after Fourier transformation, and therefore three-pulse ESEEM is easier to analyze than two-pulse ESEEM. However, in three-pulse ESEEM, some frequencies are lost depending on the interval between the first and second pulses, which is called blind spot. The blind spot appears in the frequencies around $1/\tau$.

In the case of a nucleus spin with $\boldsymbol{I} = 1$, a quadrupole interaction is also detected, which gives important information. The quadrupole interaction of the nitrogen nucleus is often analyzed for protein backbone, amino acid, metal-binding site, etc., in biological systems.

The magnetic interaction between electrons is also described in the magnetic dipole interaction. Equation (1.1.5) is rewritten for two S spins as follows.

$$\varkappa = g_e\beta_e g_e\beta_e\left[\frac{(\boldsymbol{S}\cdot\boldsymbol{S})}{r^3} - \frac{3(\boldsymbol{S}\cdot\boldsymbol{r})(\boldsymbol{S}\cdot\boldsymbol{r})}{\boldsymbol{r}^5}\right]. \tag{1.1.5}$$

In the case of electron-nucleus spin interaction, the energy levels of the electron and the nucleus are quite different, where the transitions $|\alpha_e\beta_n\rangle \Leftrightarrow |\beta_e\alpha_n\rangle$ are not included, but in the spin between electrons, the transitions $|\alpha_e\beta_e> \Leftrightarrow |\beta_e\alpha_e>$ are included. The magnitude of the dipole interaction is described as

$$D = \frac{g_e\beta_e g_e}{r^3}(3\cos^2\theta - 1). \tag{1.1.6}$$

For the detection of the electron-electron dipole interaction, some techniques have been proposed, such as PELDOR/DEER, RIDME, 2 + 1, DQC (double quantum coherence), and hole burning. The PELDOR/DEER method is the most versatile and easy to use for distance measurement in proteins.

The PELDOR/DEER method detects the dipole interaction between two electron spins using microwaves of two different frequencies and determines the distance between electron spins in the range of 10–100 Å with high accuracy. Figure 1.1.10a shows the basic three-pulse sequence for PELDOR/DEER. Supposing that a spin S1 and S2 with resonance frequencies of ω_1 and ω_2, respectively, are located in a magnetic field. First, spin S1 is detected with spin echoes with 90°–180° pulses with interval τ. After the 180° pulse for S2 is applied at the timing t between the 90°–180° pulses, there is a periodic oscillation of $\cos(2\pi Dt)$. From the oscillation pattern, the magnetic dipole interaction D between spins can be obtained. In the case of the three-pulse measurement, it is not possible to observe near $t = 0$. A four-pulse sequence is used to measure around $t = 0$ (Figure 1.1.10b). In the case of spin-coupled metal clusters, the internal exchange interaction can be obtained directly from the spin distribution. In this way, it is possible to extract various quantities depending on the measurement target, rather than simply finding the distance R.

RIDME is another technique for detecting the electron spin interaction. The technique is an intermediate method between PELDOR/DEER and hole burning. As described above, a sharp hole in the spectrum is diffused by dipole interaction. By monitoring the variation of the hole, the dipole interaction can be estimated. Figure 1.1.10c shows three-pulse RIDME sequences; the first and second pulses

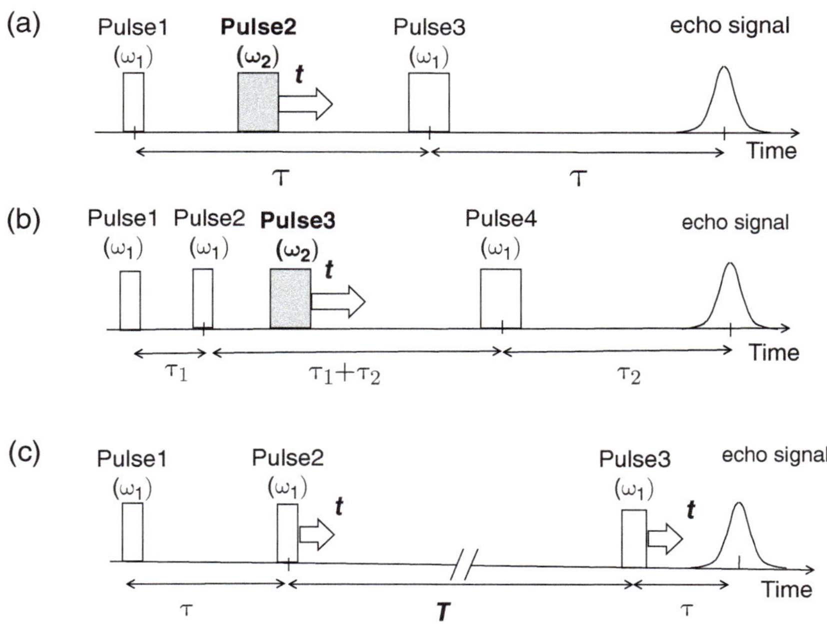

Figure 1.1.10 Pulse sequences for (a) three-pulse and (b) four-pulse PELDOR/DEER and (c) three-pulse RIDME.

($\pi/2$) invert spin to the $-z$-axis, while the third pulse detects the interaction after spectral diffusion on T. The oscillation pattern is detected by changing τ. The reason that RIDME is not so popular is simply that PELDOR/DEER is more conventional and RIDME has some difficulties. First, the nuclear ESEEM overlaps the signal. The nitrogen nucleus is difficult to distinguish from PELDOR/DEER signals, so they should be removed by subtraction of the baseline signal. If the S1 and S2 spins are the same types of magnetism, RIDME is not so great because there are no significant differences in relaxation times. However, a large effect is observed if the relaxation times of S1 and S2 are significantly different. For example, in the measurement of the interaction between the metal center and organic radical, the relaxation time differs greatly, where RIDME gives a great effect.

References

1 Carrington, A. and McLachlan, A.D. (1979). *Introduction to Magnetic Resonance*. Chapman and Hall Ltd.

2 Shweiger, A. and Jeschke, G. (2001). *Principles of Pulse Electron Paramagnetic Resonance*. Oxford University Press.

3 Slichter, C.P. (1990). *Principles of Magnetic Resonance*, vol. 1. Springer Nature.

4 Milov, A.D., Salikhov, K.M., and Shirov, M.D. (1981). Application of ELDOR in electron-spin echo for paramagnetic center space distribution in solids. *Fiz Tverd Tela (Leningrad)* 23: 975–982.

5 Dzuba, S.A., Kodera, Y., Hara, H., and Kawamori, A. (1993). The use of selective hole-burning in EPR-spectra to study spectral diffusion and dipolar broadening. *J Magn Reson Ser A* 102: 257–260.

6 Wacker, T., Sierra, G.A., and Schweiger, A. (1992). The concept of FID-detected hole-burning in pulsed EPR spectroscopy. *Isr J Chem* 32: 305–322.

7 Kulik, L.V., Dzuba, S.A., Grigoryev, I.A., and Tsvetkov, Y.D. (2001). Electron dipole-dipole interaction in ESEEM of nitroxide biradicals. *Chem Phys Lett* 343: 315–324.

1.2

Biological Applications of EPR

Isao Suetake[1,2], Risa Mutoh[3], Yuichi Mishima[2], Masatomo So[2], and Hironobu Hojo[2,]*

[1] *Department of Nutritional Sciences, Faculty of Nutritional Sciences, Nakamura Gakuen University, Fukuoka, Japan*
[2] *Institute for Protein Research, Osaka University, Yamadaoka, Suita, Osaka, Japan*
[3] *Department of Biomolecular Science, Faculty of Science, Toho University, Funabashi, Chiba, Japan*
[*] *Corresponding author*

1.2.1 Proteins and Their Structures: Domain and Intrinsically Disordered Region

Proteins are generally composed of 20 amino acids. In proteins, amino acids are connected by peptide bonds (Figure 1.2.1). The polypeptide backbone is composed of NH, carbonyl (CO), and a CHR group, in which R is the side chain of amino acids. The side chain is a specific group to individual amino acids. The sequence of amino acids in a polypeptide chain is the simplest level of protein structure, the primary structure. The locally folded structures, in which the interactions between atoms of the polypeptide backbone occur, are called the secondary structure. The most common types of secondary structures are the α-helix and the β-sheet. Both structures are held in shape by hydrogen bonds, which are formed between the carbonyl oxygen of one amino acid and the amido hydrogen of another. By associating the secondary structures, functional domains are formed. The three-dimensional domain structure has been elucidated by X-ray diffraction experiments on crystals and NMR in solution. The structures have been collected in an easily accessible protein data bank (PDB). More than one-third of proteins in eukaryotic cells, however, have disordered regions that are more than 30 amino acids in length [1]. Within disordered regions, charged amino acids are frequently observed, while the hydrophobic amino acids are rare.

Analytical Techniques for the Elucidation of Protein Function, First Edition. Edited by Isao Suetake, Rohit K. Sharma, and Hironobu Hojo.

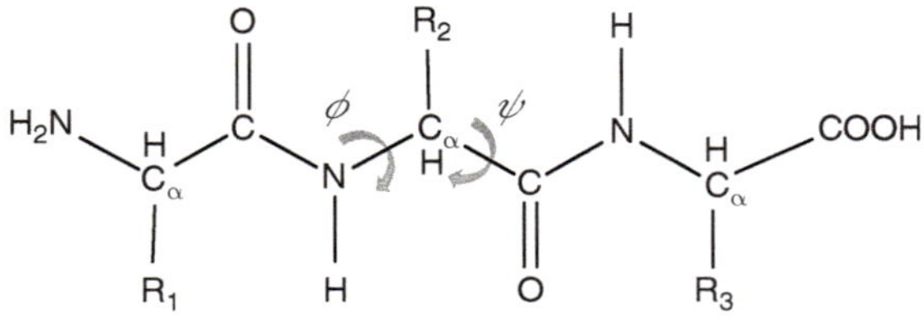

Figure 1.2.1 Peptide bond structure. A peptide is made up of an amino acid connected by peptide bonds. R (1–3) indicates side chain of amino acid. Phi (ϕ) and psi (ψ) dihedral angle rotations of the amino acid are shown.

The flexibility of these regions allows conformational change and interaction with multiple partners. EPR is an excellent technique to understand the dynamics and weak interactions.

1.2.2 Introduction of Spin Probes on Proteins

As endogenous proteins and/or protein complexes have no electron spin signal, these are usually EPR silent molecules. To analyze the movement of proteins in a site-specific manner, site-directed spin labeling (SDSL) or chemical synthesis of peptides with spin-labeled amino acid(s) is used for the introduction of a paramagnetic moiety onto a desired site. For SDSL, in general, a cysteine residue is introduced into the target site by site-directed mutagenesis, and simultaneously endogenous cysteine residue(s) is (are) replaced by Ala, Gly, and so on, to avoid structural change. Generally, the target site(s) to be spin-labeled is (are) selected to be predicted interaction site(s) from crystal structure, active site(s) determined with biochemical analyses, or not well-understood dynamic site(s). Cysteine residue possesses the unique sulfhydryl group, which is able to form a covalent disulfide bond with the spin label (Figure 1.2.2). The vast majority of spin labels are heterocyclic nitroxides with an unpaired electron in the —NO group. Nitroxide spin label for SDSL is quite small in size, similar to tryptophan residue. The commonly used chemicals carrying the nitroxide label are 4-maleimido-TEMPO (MSL), (1-oxyl-2,2,5,5-tetramethyl-D-pyrroline-3-yl) methyl methanethiosulfonate (MTSL or MTSSL), pyMTSL (also called HO-3606), and bis-MTSL (Figure 1.2.2). Compared with MSL, MTSL has a rather long linker consisting of five single bonds from Cα atom of the peptide to the 3-pyrroline ring, resulting in substantial flexibility and dynamics [2]. Due to the linker length, the expected mobility trend of MTSL is higher than that of MSL. To limit the dynamics of MTSL, the pyrroline ring of MTSL is modified to increase the tag rigidity [3]. This was successfully done by addition of a pyridyl group to position 4 of the pyrroline

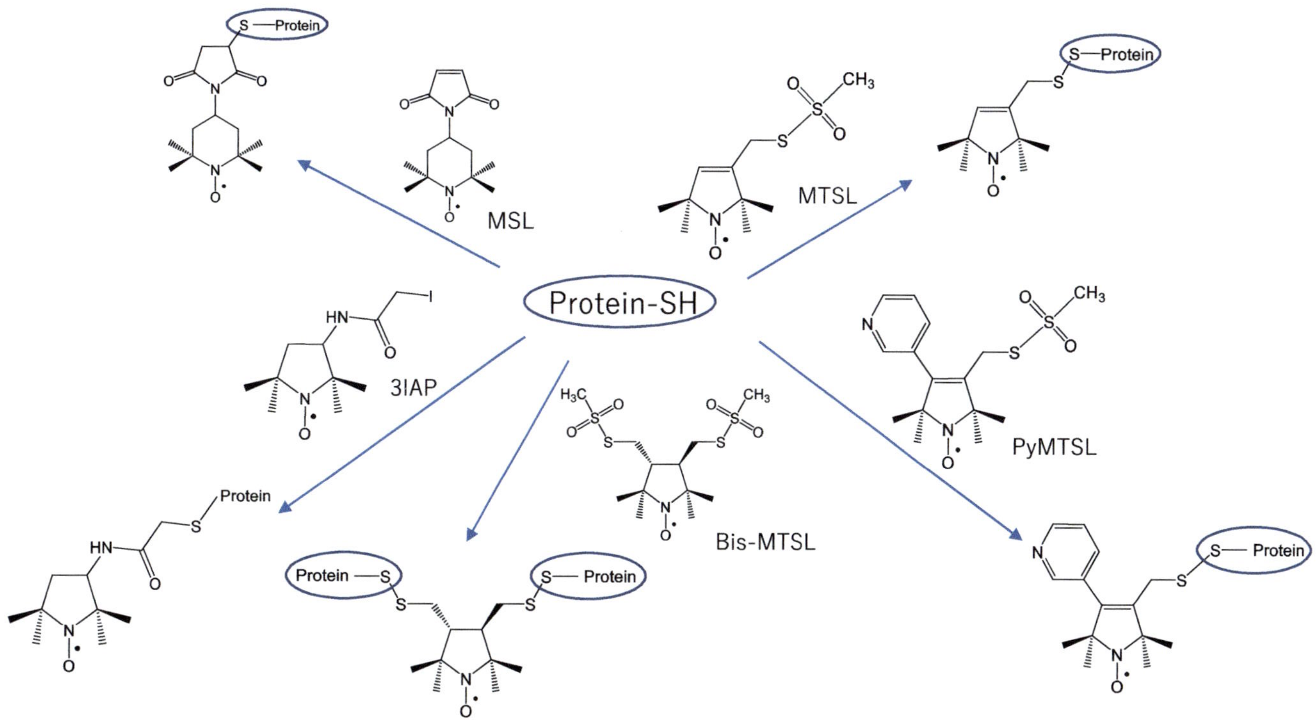

Figure 1.2.2 Molecular structures of nitroxide spin label tags on cysteine. Chemical formula of spin-labeling compounds (a maleimide and three methanesulfonothioate based tags) are shown. Chemical formula of the compounds linked to SH group of cysteine on target protein are shown.

ring resulting in 3-methanesulfonylthiomethyl-4-(pyridin-3-yl)-2,2,5,5-tetramethyl-2,5-dihydro-1*H*-pyrrol-1-yloxyl (PyMTSL) [2]. It was reported that this modification restricts spin-label movement [4].

The attachment of nitroxide spin label has been reported to have minimal effect on the overall protein structure [5–7]. As easily understandable from the chemical structure, both MTSL and PyMTSL on protein can be released with a low concentration of mercaptoethanol or dithiothreitol, allowing the labeled protein to be recovered and relabeled on the released position [8]. Recently bis-MTSL (trans-3,4-bis[[(methylsulfonyl)thio]methyl]-2,2,5,5-tetramethylpyrrolidin-1-yloxyl) has been proposed as an effective reagent for a pH dependent reversible spin labeling by insertion into a disulfide bridge [9–12].

The reagent(s) can be selected for experimental purposes among monovalent labeling reagents (MTSL, 3IAP, MSL, and PyMTSL, shown in Figure 1.2.2). As mentioned above, MTSL and 3IAP are connected to proteins with a relatively long linker (Figure 1.2.2); hence, the restriction of the dynamics of the target Cys is weaker than MSL or, in other words, the signal from the intrinsic dynamics of the labeled site on the protein is gentler compared with MSL labeling. PyMTSL possesses an additional six-member ring on MTSL, and the signal from PyMTSL has a tendency to monitor the change of hydrophobicity in the vicinity of the labeled site, compared with MTSL. Thus, depending on these chemical properties and molecular structures of nitroxide spin labels, one of labels is selected to be introduced. The labeling protocol is described in the **Notes** at the end of this section. After purification of a labeled protein, the efficiency of spin labeling is easily estimated by EPR measurement because the second integral (area under the absorption spectrum in Figure 1.2.3, bottom panel) is proportional to the spin concentration. The labeling rate of the sample can be calculated by comparing the second integral of the spin-labeled protein with that of the known concentration of spin chemicals. The protein concentration should be determined by a protein assay including a BCA assay.

In the case of low labeling efficiency, harsher reaction conditions, such as additional (second) round labeling overnight, regeneration of cysteine with a higher concentration of DTT (e.g. 10 mM) prior to the labeling, and higher concentrations of labeling reagent, could be required. In addition, the verification of whether the position of target cysteine is located in a solvent exposed state or not could be helpful for better labeling. To label cysteine located in a hydrophobic core, the labeling reaction can be performed under a denaturing condition, such as 6 M urea. For a spin reagent containing an S—S bond (such as MTSL), the addition of reductants after the labeling reaction should be avoided.

The spin labeling target is not restricted to cysteine residue. Various nitroxides to react with the phenol group of the tyrosine via a three-component

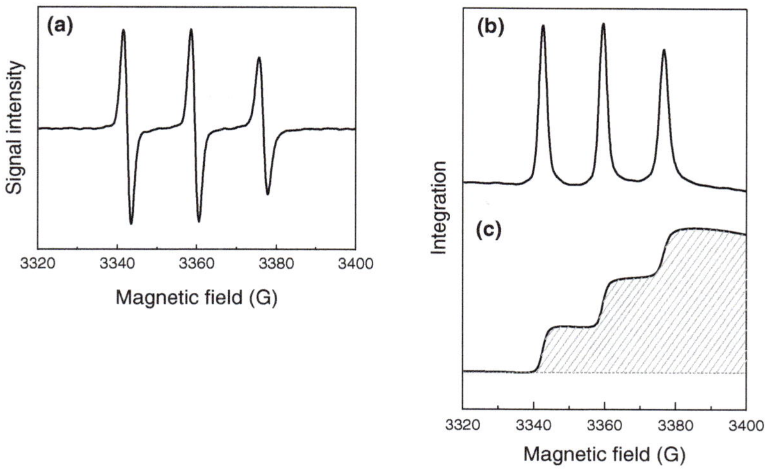

Figure 1.2.3 EPR spectrum and its integration. (a) Typical EPR spectrum of MSL. Spectrum in panel (a) is calculated from the differentiation of microwave energy absorption, shown in panel (b). For calculating the amount of spin, the EPR spectrum (a) is integrated twice (c). The amount of spin is shown by an oblique line in panel (c). In the case of MSL, the EPR spectrum is usually scanned within a range of 100 G.

Mannich-type reaction has been reported [13–15]. The isoindoline-based nitroxide: 5-amino-1,1,3,3-tetramethyl-isoindolin-2-yloxyl (Figure 1.2.4) has been described to be a good reporter for structural changes [15]. The possibility that the long reaction time of the Mannich-type reaction can modify the tryptophan and free cysteine residues has been reported [16]. However, cysteines that form an S—S bond or are pretreated with iodoacetate are not involved in the Mannich-type reaction [16].

The limitation of spin labeling by SDSL is the introduction of the spin probe to the side chain functional groups. Hence, the signal from SDSL is derived not only from the dynamics of the side chain but also from those of labeling chemicals and the main chain. In other words, the main chain dynamics is, in principle, not able to be analyzed by SDSL-EPR. In case that the spin labels on the side chain of target amino acid(s) perturb the protein function, the spin probe can be directly introduced to the main chain by chemical synthesis using 2,2,6,6-tetramethyl-N-oxyl-4-amino-4-carboxylic acid (TOAC), which is an analogue of α-amino acids (Figure 1.2.5a) [17]. In addition to TOAC, several spin-labeled amino acids have been reported. Fully synthesized and semi-synthesized proteins, which are prepared by chemical ligation between synthesized peptide and recombinant protein, can be used for EPR.

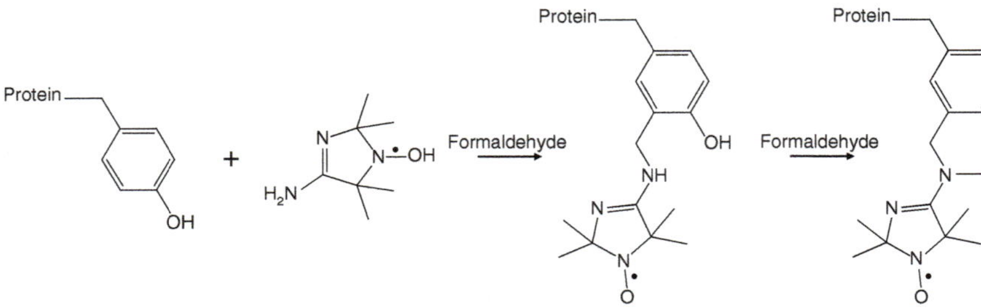

Figure 1.2.4 Reaction of one of the spin probes labeling tyrosine. Tyrosine residue is also a target for spin labeling on protein. The chemical also reacts to reduced cysteine and tryptophane, in low efficiency.

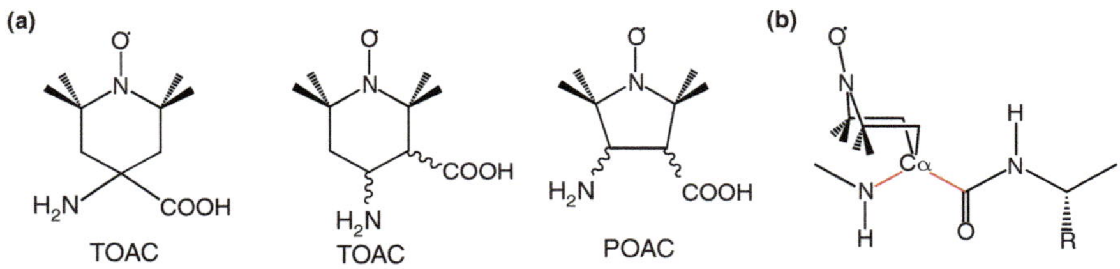

Figure 1.2.5 Spin labeling of main chain. (a) chemical formula of compounds used for spin labeling a main chain. In addition to TOAC, the chemical formula of 4-amino-1-oxyl-2,2,6,6-tetramethylpiperidine-3-carboxylic acid (β-TOAC), 3-amino-1-oxyl-2,2,5,5-tetramethyl pyrrolidine-4-carboxylic acid (POAC) are used to the labeling. (b) TOAC on peptide. The dihedral angles (ϕ and ψ) of TOAC are not freely changed (red lines). Backbone-fixed TOAC label is devoid of local mobility.

When a protein is semi-synthesized using synthetic peptide with TOAC and recombinant protein, the dynamics of the main chain in the target region can be monitored. The TOAC signal can monitor the main chain dynamics with higher resolution because the dynamics of the side chain can be neglected. The incorporation of TOAC during peptide synthesis proved to be useful for the analysis of the insertion and orientation of the target site in lipid bilayer membranes [18]. This environmental analysis is not accomplished by side chain labeling, SDSL. In the future, when semi-synthesis is prevalent among researchers, mainly in the biosciences, the dynamics of larger proteins will be better understood. However, we should remember the fact that the main chain dihedral angles (φ and ψ, shown in Figure 1.2.1) of TOAC are not freely changed (red lines in Figure 1.2.5b) and that TOAC preferentially adopts helical backbone torsion angles in crystallographic studies of synthetic peptides [19].

1.2.3 Measurement of Constant Wave (CW)-EPR Spectrum

In CW-EPR experiments, a microwave with constant frequency is applied, the applied magnetic field is swept, the absorption of microwaves is monitored, and then the derivative of the absorption is shown. The physical explanation of EPR is provided in **Section 1.1**. Among CW-EPR, several bands of EPR are used: X-band EPR (9–10 GHz) is conventionally used, while other bands (Q, W bands, and so on) can be used (Table 1.2.1). Each band ESR utilizes a different microwave frequency (Table 1.2.1). The difference of these bands is resolution. Since the resonance frequency is proportional to the magnetic field, compared with X band EPR, Q band EPR has higher sensitivity and resolution, enabling measurements with a small amount of sample. ESR with higher frequency is also suitable for analysis of anisotropic dynamics. In the case of using a nitroxide radical in X band EPR, a range of the magnetic field (100–200 G), with center ~3360 G, is typically used.

In most cases of biological EPR, nitroxide radicals including MTSL are used. EPR spectra of nitroxide radicals that contain a ^{14}N atom ($I = 1$) show split three peaks under conditions where the spin motion is high enough (Figure 1.2.6). The linewidth of EPR spectra reflects the mobility of the spins, the distance between the two spins, and the surrounding environment. In the case that a spin label is introduced to an unstructured site, a spectrum with three sharp and narrowly spaced lines is observed (Figure 1.2.6a, top). On the other hand, the spin label introduced to an immobile site shows a broad spectrum (Figure 1.2.6a, bottom). The line shape is determined by the flexibility of the segment and the steric hindrance of the residue due to tertiary structure and complex formation. After obtaining the spectrum, the movement analysis is then performed. These features are quantified by the inverse central linewidth, ΔH_0; peak intensities, V_{+1}, V_0, V_{-1}; and spectral breadth, $2T_{eff}$ (Figure 1.2.6b). The rotational correlation time, τ, is

Table 1.2.1 Frequency and wavelength of microwave used in CW-EPR.

Frequency band	Microwave frequency (GHz)
L	1
S	3.5
X	9.8
Q	34
W	94

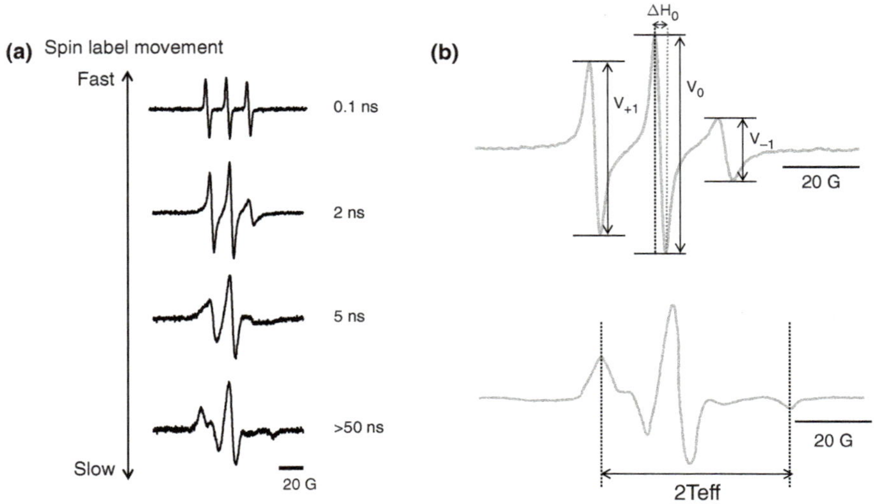

Figure 1.2.6 Representative parameters on EPR spectrum. (a) For spin-labeled sites exposed to the bulk water, the nitroxide mobility is scarcely or slightly restricted (top panel). In this case, the rotational correlation times is in the nanosecond range (τ = 0.1 ns). When the spin label side chain is restricted by interaction with neighboring side chains and/or backbone atoms, linewidths and the apparent hyperfine splitting are increased (lower panels). (b) Parameters on the EPR spectrum. ΔH_0 is the peak-to-peak linewidth of the central line, and Vm is the peak-to-peak height at low (M_I = +1), mid (M_I = 0), and high (M_I = −1) magnetic fields. A range of 2 T_{eff} is also shown.

calculated using these values. In case of fast motion ($\tau = {\sim}10^{-9}$ s), τ is calculated by the following equation: $\tau = -1.22 \times 10^{-9} \Delta H_0\{(3^{0.5}/4)(V_0/V_{+1})^{0.5} - (V_0/V_{-1})^{0.5}\}$ [20]. In contrast, in the slow-motion regime ($\tau = {\sim}10^{-9} - 2 \times 10^{-7}$ s), τ is calculated according to the following equation [21]: $\tau = 5.4 \times 10^{-10}\,(1 - T_{eff}/T_{max})^{-1.36}$, where T_{max} (= 35 G or 3.5 mT, typical for nitroxides) is the line breadth of the rigid limit for a particular spin [22]. The τ value is also calculated by other methods [23–26]), though similar values are obtained independently of the method used. In case that the spectrum contains two or more independent components with different mobility, the experimental spectrum can be more complex. When the whole spectrum is digitally subtracted either by the typical fast- or slow-component dominant spectrum (which is usually obtained from the control experiment), the spectrum caused by the slow or fast component is respectively obtained. It can also be analyzed by assuming several independent motions. Instead of spectral subtraction, the obtained spectra can be fitted using simulated spectra. Since the line shapes of EPR spectra are well understood, a simulation spectrum containing components with different motion can be obtained using an application such as easy spin [27].

For evaluating the measured dynamics, the calculated τ value can be compared with the theoretical τ in isotropic motion of a spherical rigid body protein in water, which is calculated by the formula MW/2.4, in which MW is the molecular weight [28].

The observed apparent rotational motion of the spin label on the side chain of protein is the sum of rigid body rotational diffusion of the protein and motion of the spin-labeled site. The apparent rotational motion is calculated by the equation $1/\tau = 1/\tau_P + 1/\tau_L$, where τ is the apparent experimental correlation time, and τ_P and τ_L are true correlation times of the protein and spin label, respectively. When τ_P is assumed to be proportional to viscosity (for example, the measurements are performed in the presence of 0, 20, and 50% glycerol [29]) and τ_L is assumed constant, $1/\tau_L$ is estimated by extrapolating the $1/\tau$ vs 1/viscosity plot to 1/viscosity = 0, and then the true value of τ_p is calculated as $1/(1/\tau - 1/\tau_L)$ at 0% glycerol.

CW-EPR can also provide the information of inter-spin distances up to 20 Å. When the distance between two spin labels is close, spin-spin interaction and spin exchange can be observed, and the line shape changes. Since the relationship between distance and exchange rate is well understood, the distance between two spin labels can be estimated accurately. To obtain the distance between two segments, two spin labels need to be introduced. In some special cases such as high-density conditions, as in crystals and amyloid fibrils (see Section 1.2.4.2), a single label can also be applied.

The SDSL-EPR technique is applied not only for analyzing protein dynamics but also the environment of a protein. The spectrum of SDSL-EPR also shows the accessibility toward paramagnetic quencher molecules, spin label microenvironment (polarity), and transient structural changes. By using quencher [Ni (II) ethylenediaminediacetate (NiEDDA), Cr (III) oxalate (CrOx), O_2 and so on], the localization of spin (lipid bilayer, aqueous phase, and protein interior) is discriminated [30–32]. By analysis of the collision frequency with paramagnetic reagents, a remarkable structural information is obtained; a specific region of the T4 lysozyme shows characteristics of α-helix and β-sheet [33]. This allows us to obtain structural information of proteins, for which NMR and/or X-ray crystallography is not easily applied.

1.2.4 Application of CW-EPR to Protein (Clock Protein, Amyloid Proteins, and HP1)

CW-ESR has been utilized to analyze protein dynamics and local structural changes of a target protein depending on its environment. The technique is also used to detect protein-protein and protein-lipid interactions. In this section, several examples of the application of CW-EPR on protein research are described.

1.2.4.1 Clock Proteins

Circadian rhythms are physical, metabolic, and behavioral changes with a 24-hour period and are observed ubiquitously from prokaryotes to eukaryotes. Cyanobacteria are the simplest organisms that show circadian rhythms. Cyanobacterial circadian clock machinery is composed of three clock proteins, called KaiA, KaiB, and KaiC, and known to be reconstituted in vitro. These Kai proteins periodically associate and dissociate each other and form various complexes within 24 h [34]. Structural biology is essential for understanding the regulatory mechanism of circadian clocks. The crystal structures of each Kai protein have been solved [35–37]. The structure of Kai complexes using domain mutants have been solved by X-ray crystallography [38], cryo-electron microscopy analysis [39], and solution NMR analysis [38]. The interaction between KaiB-KaiC [40] and KaiA-KaiC [41] has also been detected by solution NMR. Recently, the whole KaiABC-complex structure was reported using small-angle neutron scattering (SANS; Chapter 2) [42], but this technique does not reveal the detailed binding site. However, these methods require a stable complex formation since weak interactions are hard to detect. In this regard, the circadian clock machinery of cyanobacteria is one model system for the unstable interaction research because complex components change through weak interaction over time. Although the complexes of KaiA-KaiC and KaiB-KaiC were detected by several methods (gel filtration chromatography, native PAGE, etc.), a stable KaiA-KaiB complex has not been detected. By using SDSL-EPR, the KaiA-KaiB interaction is clearly demonstrated [26]. When KaiB was spin-labeled via non-native Cys residue (T64C) with MTSL (Figure 1.2.7a), in the absence of KaiA the EPR spectrum of MTSL-KaiB$_{T64C}$ was broad. After incubation for 3 h with KaiA, the spectrum of MTSL-KaiB$_{T64C}$ showed narrower EPR peaks compared to those in the absence of KaiA (Figure 1.2.7b). In addition, the subtraction of the spectrum before incubation from that after incubation in the presence of KaiA showed extremely narrow peaks corresponding to free MTSL (Figure 1.2.7c), suggesting that the spin labels were released from MTSL-KaiB$_{T64C}$ by the addition of KaiA (Figure 1.2.7b). This suggests that disulfide bond-exchanging reactions between MTSL-KaiB$_{T64C}$ and KaiA occur and KaiB$_{T64C}$-KaiA dimer forms.

The existence of the KaiB$_{T64C}$-KaiA dimer was also demonstrated by SDS-PAGE in the presence of C-terminal domain of KaiA, which has only one cysteine, C272, near the functional residue, H270, suggesting that the disulfide bond is replaced with one between KaiB$_{T64C}$ and C272 of KaiA. These phenomena were observed only for MTSL-KaiB$_{T64C}$ incubated with KaiA but not for MTSL-KaiB$_{A101C}$, which is located on the opposite side of the molecule from T64 in the crystal structure, or at a low temperature at which MTSL-KaiB$_{T64C}$ scarcely reacts with KaiA, suggesting that conformational changes around T64 at physiological temperature are required for the interaction. As well as the KaiA-KaiB interaction described above, KaiC interaction sites on KaiA were determined by SDSL-EPR [43].

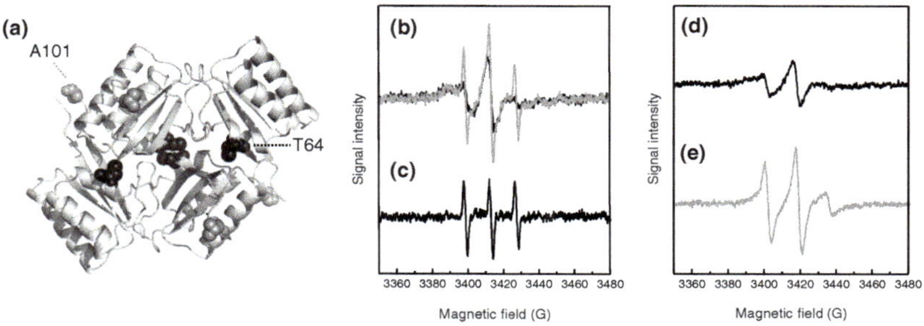

Figure 1.2.7 Structure of KaiB and EPR spectra of MTSL-KaiB$_{T64C}$ and MSL-KaiB$_{T64C}$. (a) Crystal structure of KaiB. The labeling sites are shown as black spheres (T64) and gray spheres (A101). (b) EPR spectra of MTSL-KaiB$_{T64C}$ in the presence of KaiA before (black) and after (gray) incubation for 3 h. The subtraction spectrum from gray spectrum to black spectrum is shown in panel (c). (d) and (e) EPR spectra of MSL-KaiB$_{T64C}$ (d) before and (e) after incubation. The figure is taken and modified from Mutoh et al. [26, 44].

The EPR spectrum also determines whether the target amino acids are located on the surface or inside of the protein. The EPR spectrum of MTSL-KaiB$_{T64C}$ showed the broad peaks compared to those of MTSL-KaiB$_{A101C}$, which is consistent with the fact that T64 of KaiB is buried inside the molecule whereas A101 is exposed (Figure 1.2.7a) [26]. Local structural changes can also be analyzed by EPR with other spin labeling. Bulky spin label group, MSL (Figure 1.2.2), is more sensitive to steric hindrance due to surrounding amino acid residues than MTSL (Figure 1.2.2). When purified MSL-KaiB$_{T64C}$ was incubated alone for 24 h, the rotational correlation time, τ, was significantly decreased from 79 to 1.9 ns with a large spectral change (Figures 1.2.7d and 1.2.7e) [44]. This observation cannot be performed by MTSL.

The line shape significantly depends on steric hindrance. When one of the interaction sites of KaiB (N19) upon binding of KaiC was labeled with 3IAP (see Figure 1.2.2, 3IAP-KaiB$_{N19C}$) [45], EPR spectra of 3IAP-KaiB$_{N19C}$ in the absence of phosphomimic KaiC mutant (KaiC$_{EE}$) showed narrow peaks. When KaiC$_{EE}$, whose affinity for KaiB is higher than KaiC$_{WT}$, was added in the reaction, broader EPR spectra containing a slow motional component were observed. This means that the vicinity of the spin label was covered by KaiC$_{EE}$, resulting in the restriction of the spin label movement.

1.2.4.2 Amyloid Proteins (Aβ Peptide, β_2-microglobulin, α-synuclein, Tau, and Prion)

Highly ordered protein aggregates, "amyloid," –related diseases are called amyloidoses, and there have been over 30 diseases reported, including systemic and localized amyloidoses [46, 47]. The number of patients of some of them, such as type 2 diabetes, in which islet amyloid polypeptides form amyloid fibrils, and Alzheimer's disease, in which amyloid β peptides (Aβ) form fibrils, is increasing, and it has become a social problem. In order to understand the properties and effects of amyloid fibrils, their molecular structures and changes are important. However, standard experimental approaches to structural analysis, including X-ray crystallography and solution NMR, have been limited, because amyloid fibrils are inherently insoluble, high molecular weight assemblies. Due to the difficulty applying other methods, solid-state NMR spectroscopy has been used for structure determination of amyloid fibrils.

For the studies of amyloid fibrils, EPR is also effective for the measurements of insoluble materials. SDSL combined with EPR spectroscopy allows the measurement of the backbone distances and mobility of spin-labeled position in amyloid fibrils. One or more cysteine mutations were introduced and spin-labeled into amyloid proteins, and the mobility around the region and the intra- and inter-molecular distances were measured to clarify the molecular structure. From the line width of the EPR, the motility around the spin label can be known, and it is possible to estimate the orientation of the amino acid residue in the amyloid fibrils. When the residue is located in the core region within amyloid fibrils, line broadening occurs due to reduced motility. When located outside the fibril core, the spin label moves fast, and the EPR spectrum is sharp and has a narrow linewidth, as mentioned above (Section 2). Aβ peptide is a 40- or 42-residue peptide ($A\beta_{40}$ and $A\beta_{42}$) produced by abnormal cleavage from the amyloid precursor protein, APP, which is extremely toxic and forms amyloid fibrils. The structural features of amyloid fibrils of Aβ were revealed by using EPR [48]. From the linewidth analysis using SDSL-EPR, 10 N-terminal residues and 2 ($A\beta_{40}$) or 4 ($A\beta_{42}$) C-terminal residues were more flexible in the amyloid fibrils of both peptides [48]. In addition, the central residues (23–29) were also highly flexible, suggesting that the β-strand core region bends in this region to form a U-shaped structure. EPR data were consistent with the overall structure of Aβ fibrils revealed by solid-state NMR [49, 50]. A similar study was also performed on β_2-microglobulin amyloid fibrils, which are causative of dialysis-related amyloidosis. The 19–91 region of 100 amino acid residues showed a low degree of motion, indicating the formation of the fibril core. Thus, the mobility of a spin label within amyloid fibrils gives an important information of amyloid fibril cores and has been applied to various proteins.

Spin-spin interactions have also been widely applied in the study of amyloid fibrils. Amyloid fibrils are composed of cross-β structures, and their classification is roughly divided into parallel in-register, parallel out-of-register, anti-parallel in-register, and anti-parallel out-of-register (Figure 1.2.8). It remains controversial which arrangement each amyloid fibril has, and it is also an interesting question in terms of protein structure. The CW-EPR measurement, which is very sensitive to the distance of 4–20 Å, is suitable for the analysis of the cross-β

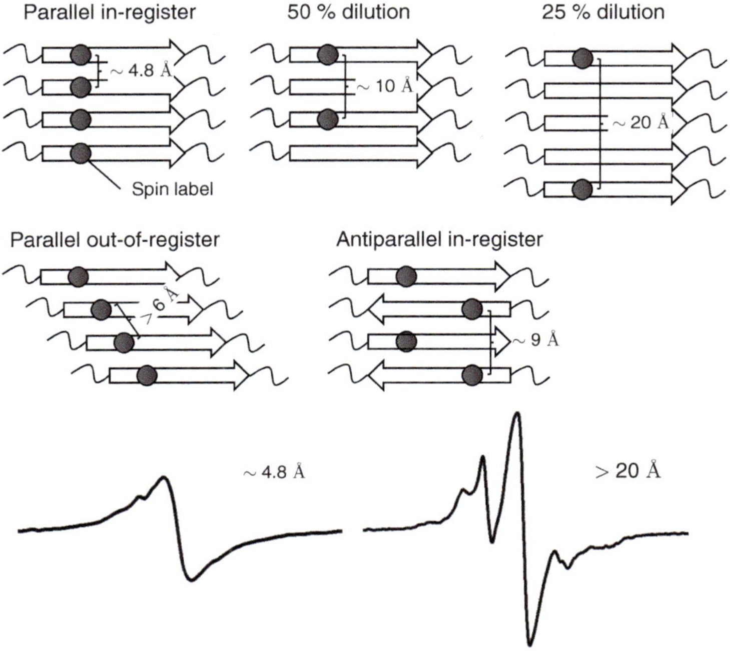

Figure 1.2.8 Structure models of amyloid cross-β sheets and expected EPR spectra. Spin label is introduced to one residue. The distance between spin labels depends on the amyloid structure and dilution of the spin label. EPR spectra depend on the inter-spin distance. At ~4.8 Å corresponding to parallel in-register (100 % labeled), a single peak due to spin exchange should be observed. Broaden spectra with three peaks of the nitroxide radical are observed depending on structure models and/or spin dilution. Scan width is 150 G.

structure with a 4.8 Å width and has answered the above question. In a parallel in-register arrangement of β structure in which spin labels are stacked in-line perpendicular to the fibril axis, a single-line EPR spectrum can be observed due to spin exchange between multiple spin labels (Figure 1.2.8, bottom left). Mixing spin-labeled proteins and non-labeled proteins to create fibrils reduces the probability that the labeled proteins are next to each other, increases the distance between spin labels, and suppresses the exchange, leading to broad spectra with three peaks (Figure 1.2.8, bottom right). The distance can be estimated from the dilution rate and the line shape, and the arrangement of β-strands can be estimated. In the case of Aβ fibrils, the 50% diluted spectrum had a significantly sharper linewidth than the 100% labeled fibrils [48]. Since the relationship between line broadening due to dipole interaction and the inter-spin distance is well understood, the inter-strand distance was calculated from the fitting of each inter-spin distance in the dilution series. Since the distance between labels for amino acids of 14–38 residues in the core region was 9–11 Å, it was found that the structure was a parallel in-register structure in which the same amino acid residues are stacked on top of each other along the fibril axis. Similarly, it was revealed that amyloid fibrils such as α-synuclein, tau, and prion protein also have a parallel in-register structure [51–53]. Recently, the overall structure of amyloid fibrils has been reported in large numbers by cryo-EM technique, which has been dramatically developed in recent years. All of these structures formed a parallel in-register structure consistent with the EPR results.

1.2.4.3 Heterochromatin Protein 1 (HP1)

The genome's primary function is to store, propagate, and regulate the expression of genetic information required by an organism to function. In addition, there exists another modulation, or regulation, mechanism of gene expression called epigenetics, which is independent of changes in the DNA sequence. Under epigenetic regulation, (1) expression of genetic information is changed without changing the nucleotide sequence, and (2) memory of the change is propagated to the next generation. Major molecular bases underlying epigenetics are DNA methylation, histone post-translational modifications, noncoding RNA, and so on [54]. DNA methylation, histones modifications, and noncoding RNA affect the higher-order structure of chromatin, one of whose basic structures is the nucleosome. Changes in the higher-order structure of chromatin are thought to be one of contributors for gene expression regulation by epigenetic mechanisms.

In the nucleus, both electron-dense and -sparse regions exist under electron microscope. The electron-dense region is called heterochromatin, which is inactive in gene expression, and the electron-sparse region is called euchromatin, most of which is genetically active. Single nucleosome unitcomprises two histone dimers of H2A-H2B and one histone tetramer of $(H3\text{-}H4)_2$, wrapped around by 145 bp DNA. A variety of histone modifications, which are the major epigenetic players, exist in all eukaryotes. Lysine residue is modified with acetylation, methylation, ubiquitylation, sumoylation, lactation, etc.; arginine residue with methylation, threonine, serine, tyrosine residues with phosphorylation; and glutamate with ADP-ribosylation [55–59]. Most of the modifications are observed on the tails of core histones, which are protruding from the core part of nucleosomes and do not form rigid secondary structures (Figure 1.2.9).

Using yeast as the material and chromatin immunoprecipitation (ChIP) as the technique, it was reported that histone H3, of which K9 residue is methylated (H3K9me), is localized to pericentromeric heterochromatin with Swi6 (fission yeast HP1 orthologue), and H3K4me at euchromatin [60]. According to technological progress, such as increasing available specific antibodies and sequencing technique (ChIP-sequencing), many genome-wide analyses of the modification

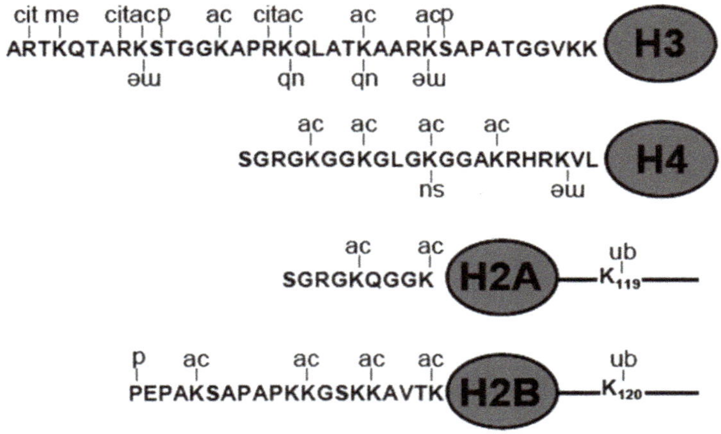

Figure 1.2.9 Some modifications on tails of histones; acetylation (ac), methylation (me), ubiqutination (Ub), phosphorylation (P), sumoylation (su), citrullination (cit).

positions have been identified [61, 62]. Methylation modification is position specific and functions to both promote and suppress gene expression, while acetylation modification occurs on several different lysine residues at the same occasion and generally promotes gene expression. In general, lysine residue accepts three types of methylation: mono- (me1), di- (me2), or tri- (me3) methylation. H3K9me1 and me2 are localized to euchromatin and H3K9me3 to constitutive heterochromatin regions [63].

In mammals, the HP1 family comprises three variants, HP1α, HP1β, and HP1γ. All the HP1s contain a conserved domain called chromodomain (CD), which is responsible in specific recognition of H3K9me3, and a chromoshadow domain (CSD), which is either homo- or hetero-dimerized (Figure 1.2.10a) [64]. All HP1 isoforms possess an intrinsically disordered region (IDR) at N-terminus (N-tail), and C-terminal region (C-tail), and at hinge region (HR) linking CD and CSD. Different from HP1α and HP1β, which localizes to heterochromatin regions, HP1γ is reported to be localized to euchromatin regions [65]. However, all types of HP1 show similar affinity to the H3K9me3 peptide. Interestingly, toward the reconstituted nucleosomes, isoform-specific characteristics have been reported; not HP1γ but HP1α selectively bounds to H3K9me3-containing nucleosomes [66]. This result partly explains the distinct localization of HP1α and HP1γ in the nucleus. In addition, HP1α does not bind to mononucleosomes without linker DNA but to mononucleosomes with linker DNA and to dinucleosomes in higher affinity (Figure 1.2.10b, [67]). Different from biochemical analysis, the dynamics of HP1 still remain to be solved. MD simulation shows that single CD in dimer and CSD dimer and IDR dynamically move on a dinucleosome, and even another

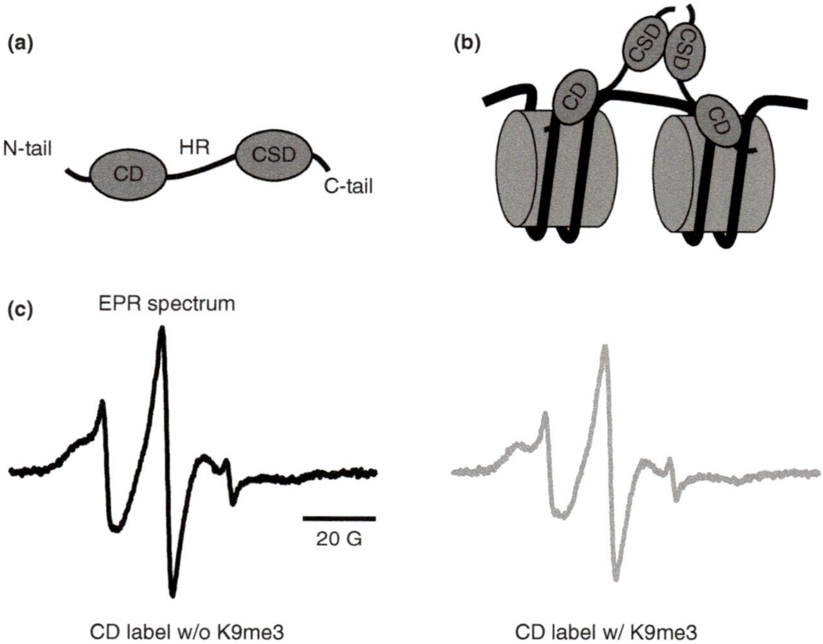

Figure 1.2.10 Schematic illustration of an HP1 molecule and EPR spectrum. (a) Schematic illustration of HP1. (b) HP1 dimer on dinucleosome. (c) EPR spectrum of CD in full-length HP1a. Spectrum of MSL-C59 in the absence (left) and presence (right) of H3K9me3 peptide.

CD of dimer is fixed on one of two nucleosomes [68]. The dynamics of HP1 recently have started to be elucidated.

By structural analysis, CD has been reported to interact with the H3K9m3 peptide via aromatic cage [69]. As expected, the dynamics of CD within full-length HP1α (in either dimer or monomer) in the presence of H3K9me3 is ~1.5-fold lower than those in its absence [29]. In contrast, only a slight or no immobilizing effect of H3 (1–30) K9me3 on the truncated CD domain is observed [29]. Together, CD dynamics in the full-length HP1 are dampened by connecting to other regions at the distal ends, irrespective of a monomeric or dimeric structure. By EPR analysis, species-specific HP1 regulation is also proposed. In full-length HP1α and HP1γ, the rotation correlation time of CD in the absence of H3K9me peptide is lower than that in the presence of the peptide (Figure 1.2.10c) [29]. This is unlike Swi6, in which CD-CD interaction is observed in an inactivated state, in the absence of the H3K9me peptide [70]. This is due to the property of Swi6, different from human HP1. Swi6 possesses a loop region in each CD, whose sequence is similar to the histone H3 tail surrounding K9, and the region mutually binds to another CD in dimer to form a closed state [70]. When Swi6 in the inactivated state encounters the methylated histone H3 or nucleosomes, the intermolecular interaction via loop-CD is disrupted, and Swi6 becomes open state. The dynamics of human HP1 CD constantly allows a Brownian search for H3K9me3, which is partly limited by HR and CSD. Compared with CD and CSD, N-tail, HR, and C-tail show fast dynamic movement [71]. Future EPR spectroscopy can reveal the dynamic regulation of epigenetics.

1.2.5 Measurement of Longer Distance between Spin-spin (HP1, Tau, α-synuclein)

CW-EPR and pulsed-EPR such as double electron-electron resonance (DEER) are used for the measurement of inter-spin distances. Paramagnetic relaxation enhancement (PRE)-NMR provides both protein structure and dynamics information up to 25 Å. The measurement of the distance between two placed spin labels by CW-EPR is within the range of 8–20 Å, and that by DEER is up to ~80Å. For easier understanding of the distance, it can be compared with the diameter of a spherical protein of 60 kDa (around 50 Å). DEER, which determines the distance of two points on a protein or protein complex (see Section 1.1), has a great advantage to analyze the conformational changes or molecular flexibility. In case that distinct several sharp peaks are observed, stepwise movements can be expected. No peaks indicate the positions randomly move. Its validity is clear by the following applications, HP1 and amyloid proteins.

The CSD region of HP1, which is dimerized, has been crystalized, and its structure has been solved (pdb; 3I3C). However, the dynamics of full-length dimer in water has not been clearly understood, i.e. rigid or fluctuated. As the molecular weight of CSD is approximately 7 kDa, DEER analysis can be applied to the dimerized CSD in full-length HP1. The distribution of the distance between two unpaired electrons in spin introduced on endogenous C160 in HP1α dimer shows a single sharp peak (representing more than 90%), with a length of 2.6 nm. The distance is well matched to the one calculated from crystal structure. Therefore, even in water, CSD dimer structure in full-length HP1α is merely fluctuated. Similar to HP1α, CSD of full-length HP1γ forms a stable dimer with fixed position. When stepwise distances are observed using DEER, the detailed movement of domains or proteins will be unraveled with cryo-electron microscopy technique.

DEER analysis has also been applied to amyloid proteins. Although detection of intermediates of amyloid formation and clarifying their roles are important not only for the treatment and prevention of diseases but also for understanding the physical property of proteins, it is difficult to obtain information of the transient sates at several nanometers. Han and coworkers investigated the process of tau aggregation using DEER measurements [72]. It has been known that the microtubule binding repeat in the C-terminal region of a human tau isoform, 2N4R, which contains four repeat domains, is important for aggregation. Among this region, the hydrophobic regions called paired helical filaments (PHFs) have been considered to be important amyloid core regions. Two spin labels were introduced at both ends (G272/S285 or G303/S316) of the amyloidogenic hexapeptides, PHF6 (306–311) or PHF6* (275–280), of the C-terminal fragment of tau (255–441). The distributions of intramolecular spin-label distances were measured using DEER. Tau filaments start to form by the addition of heparin. Before the addition of heparin, the distances between spin labels in monomeric tau were 3.38 and 3.16 nm for PHF6 and PHF6*, respectively. After the addition of heparin, the distances increased to 4.1 and 4.35 nm, indicating that tau formed more extended conformation. This extension occurs right after the addition of heparin and essentially completes within ~10 minutes of adding heparin to the tau sample, which is unexpectedly rapid considering the fibril formation kinetics reaches a plateau only after 12–16 hours of aggregation time. Furthermore, it was revealed, through DEER experiments, that cofactors similar to heparin exist in the brain and promote the formation of fibrils whose structure is different from heparin-induced fibrils.

Regarding α-synuclein, conformational changes of monomer in the presence of amyloid fibrils are investigated by DEER experiments [73]. In amyloid fibrils, the center NAC region of α-synuclein forms a fibril core, while N- and C-terminal regions are flexible. α-synuclein, which is an intrinsically disordered protein, generally takes random coil structures before amyloid formation, while the dynamic interactions between molecules exist and play important roles in amyloid formation. Kumari et al. measured the distance between two spin labels introduced at positions E20C and E35C of α-synuclein and investigated conformational changes of α-synuclein monomers interacting with fibril seeds [73]. In a monomeric state before amyloid formation, the distances between two spin labels were mainly populated at ~2.2 nm, whereas that after the addition of fibril seeds resulted in the increase of a long distance, ~3 nm in addition to the monomer distance, ~2.3 nm. Combined with NMR data, this result demonstrates that α-synuclein-fibrils interactions weakened the intramolecular contacts, favoring more extended conformations, resulting in amyloid growth.

Recent developments of cryo-EM and NMR have become popular methods for structural analysis of amyloid fibrils at atomic resolution and observation of conformational dynamics of prefibrillar states at amino acid resolution. On the other hand, the wide range of applicable sample conditions of EPR allow to detect conformational changes in fibril formation processes. Thus, DEER analysis can clarify the detailed movement of domains and the aggregate structure of protein in the future.

1.2.6 Biophysical Functions of Protein Dynamics

Biomolecular liquid-liquid phase separation (LLPS) called "membraneless organelles" is the spontaneous condensed compartment without lipid membranes. LLPS are mainly composed of proteins and/or nucleic acids and able to form and disperse reversibly, unlike solid aggregation. This phenomenon is related to various important biological functions including replication and translation, stress response, and homeostasis. LLPS is also related to diseases such as neurodegenerative diseases, suggesting that it controls biological phenomena. However, the intrinsically disordered nature of related proteins and their wide dynamic range make it challenging to apply various structural analysis methods. EPR has revealed the protein mobility and conformational changes in LLPS. Liu et al. demonstrated that the tau protein has very high motility even during LLPS and has properties as a liquid [74]. EPR experiments also revealed the mobility and structure of RNA-binding protein FUS related to amyotrophic lateral sclerosis (ALS) during LLPS. CW-EPR line shapes and DEER measurements of intramolecular distances revealed reduced mobility and a compaction of FUS in the condensed phase [75].

Human HP1α also undergoes LLPS in vitro, when it is phosphorylated and in combination with DNA [76]. Drosophila HP1 protein, HP1a, also forms phase-separated condensates in vivo [77]. In contrast, human HP paralogues, HP1β and γ, cannot undergo LLPS by themselves or with DNA in vitro, [76, 78, 79]. HP1α-DNA condensates are resistant to mechanical disruption by large forces, while HP1β can be readily dissolved [78]. In addition, HR of HP1α is sufficient to DNA compaction and LLPS with DNA, and N- and C-terminal disordered positively and negatively regulate the LLPS formation activity of HR, respectively [78]. In living cells, however, HP1 has only a weak capacity to form liquid droplets, and heterochromatin foci lack a separated liquid HP1 pool [80]. Thus, these in vitro biochemical properties do not simply translate to in vivo phenomena, at present. Study of protein dynamics could interconnect the discrepancy, by future investigations connecting in vitro and in vivo.

1.2.7 Summary/Conclusion

As described, EPR methods including CW- and pulsed-EPR are useful techniques to study the structural change and function of proteins. EPR provides us with various measurements including mobility, distance, accessibility, and orientation of the spin-labeled molecule. However, unlike NMR, EPR is not familiar to bioscience researchers at the moment, due to several reasons, including that spin-labeling is required for the detection, the EPR apparatus is expensive, and that the outstanding experimental protocols and gold standard analysis method have not been widely shared. EPR spectra contain various spin characteristics including mobility, interaction, and anisotropy, making it complex, especially in complex systems such as biological systems consisting of various molecules including membranes and cytoskeletons. In the future, with general/helpful interface to users, CW-EPR and DEER can be useful apparatuses, being easily used by bioscientists, coupling with developing automatic systems, such as AI.

Notes

To efficiently perform the labeling reaction with the chemicals containing spin moiety, it is important that the side chain of cysteine is firstly reduced with reductants such as dithiothreitol (DTT) and then the reducing reagents are thoroughly removed before the addition of spin label chemical to avoid the reduction of nitroxide group and to inhibit conjugation with the reducing agents. A desalting column, such as Hi-Trap desalting column or PD10 column (Cytiva), is the easiest and most reliable way to eliminate the reducing reagents from sample solution. Immediately after desalting, 2 molar excess of the nitroxide spin label over cysteine (i.e. MSL or MTSL stock at 0.1 M in dimethyl formamide [DMF]) is added in a dropwise manner with quick and gentle mixing and then is incubated under appropriate conditions (e.g. at 20 °C for 3 h or at 4 °C for O/N with gentle rocking). The incubation conditions depend on protein stability and degree of exposure of the cysteine residue. In the reaction mixture, no more than 1% (v/v) DMF is generally used for maintaining protein structure. After labeling reaction, unreacted labeling chemicals should be removed by desalting column or dialysis, in order not to contaminate the signal derived from free probe.

References

1 Ward, J.J., Sodhi, J.S., McGuffin, L.J. et al. (2004). "Prediction and functional analysis of native disorder in proteins from the three kingdoms of life." *J Mol Biol* 337 (3): 635–645. doi: 10.1016/j.jmb.2004.02.002.

2 Schilder, J., Liu, W.M., Kumar, P. et al. (2016). "Protein docking using an ensemble of spin labels optimized by intra-molecular paramagnetic relaxation enhancement." *Phys Chem Chem Phys* 18 (8): 5729–5742. doi: 10.1039/c5cp03781f.

3 Jeschke, G. (2013). "Conformational dynamics and distribution of nitroxide spin labels." *Prog Nucl Magn Reson Spectrosc* 72: 42–60. doi: 10.1016/j.pnmrs.2013.03.001.

4 Fawzi, N.L., Fleissner, M.R., Anthis, N.J. et al. (2011). "A rigid disulfide-linked nitroxide side chain simplifies the quantitative analysis of PRE data." *J Biomol NMR* 51 (1–2): 105–114. doi: 10.1007/s10858-011-9545-x.

5 Brown, L.J., Sale, K.L., Hills, R. et al. (2002). "Structure of the inhibitory region of troponin by site directed spin labeling electron paramagnetic resonance." *Proc Natl Acad Sci U S A* 99 (20): 12765–12770. doi: 10.1073/pnas.202477399.

6 Langen, R., Oh, K.J., Cascio, D. et al. (2000). "Crystal structures of spin labeled T4 lysozyme mutants: Implications for the interpretation of EPR spectra in terms of structure." *Biochemistry* 39 (29): 8396–8405. doi: 10.1021/bi000604f.

7 Perozo, E., Cortes, D.M., Sompornpisut, P. et al. (2002). "Open channel structure of MscL and the gating mechanism of mechanosensitive channels." *Nature* 418 (6901): 942–948. doi: 10.1038/nature00992.

8 Berliner, L.J., Grunwald, J., Hankovszky, H.O. et al. (1982). "A novel reversible thiol-specific spin label: Papain active site labeling and inhibition." *Anal Biochem* 119 (2): 450–455. doi: 10.1016/0003-2697(82)90612-1.

9 Berliner, L.J. (2012). "History of the use of nitroxides (aminoxyl radicals) in biochemistry: Past, present and future of spin label and probe method." *Nitroxides - Theory, Experiment and Applications*. doi: 10.5772/2887.

10 Lösel, R.M., Philipp, R., Kálai, T. et al. (1999). "Synthesis and application of novel bifunctional spin labels." *Bioconjug Chem* 10 (4): 578–582. doi: 10.1021/bc980138v.

11 Savich, Y., Binder, B.P., Thompson, A.R. et al. (2019). "Myosin lever arm orientation in muscle determined with high angular resolution using bifunctional spin labels." *J Gen Physiol* 151 (8): 1007–1016. doi: 10.1085/jgp.201812210.

12 Stevens, M.A., McKay, J.E., Robinson, J.L. et al. (2016). "The use of the Rx spin label in orientation measurement on proteins, by EPR." *Phys Chem Chem Phys* 18 (8): 5799–5806. doi: 10.1039/c5cp04753f.

13 Le Breton, N., Martinho, M., Mileo, E. et al. (2015). "Exploring intrinsically disordered proteins using site-directed spin labeling electron paramagnetic resonance spectroscopy." *Front Mol Biosci* 2: 21. doi: 10.3389/fmolb.2015.00021.

14 Lorenzi, M., Puppo, C., Lebrun, R. et al. (2011). "Tyrosine-targeted spin labeling and EPR spectroscopy: An alternative strategy for studying structural transitions in proteins." *Angew Chem Int Ed Engl* 50 (39): 9108–9111. doi: 10.1002/anie.201102539.

15 Mileo, E., Etienne, E., Martinho, M. et al. (2013). "Enlarging the panoply of site-directed spin labeling electron paramagnetic resonance (SDSL-EPR): Sensitive and selective spin-labeling of tyrosine using an isoindoline-based nitroxide." *Bioconjug Chem* 24 (6): 1110–1117. doi: 10.1021/bc4000542.

16 McFarland, J.M., Joshi, N.S., and Francis, M.B. (2008). "Characterization of a three-component coupling reaction on proteins by isotopic labeling and nuclear magnetic resonance spectroscopy." *J Am Chem Soc* 130 (24): 7639–7644. doi: 10.1021/ja710927q.

17 Marsh, D., Jost, M., Peggion, C. et al. (2007). "TOAC spin labels in the backbone of alamethicin: EPR studies in lipid membranes." *Biophys J* 92 (2): 473–481. doi: 10.1529/biophysj.106.092775.

18 Schreier, S., Bozelli, J.C., Jr., Marín, N. et al. (2012). "The spin label amino acid TOAC and its uses in studies of peptides: Chemical, physicochemical, spectroscopic, and conformational aspects." *Biophys Rev* 4 (1): 45–66. doi: 10.1007/s12551-011-0064-5.

19 Toniolo, C., Valente, E., Formaggio, F. et al. (1995). "Synthesis and conformational studies of peptides containing TOAC, a spin-labelled C alpha, alpha-disubstituted glycine." *J Pept Sci* 1 (1): 45–57. doi: 10.1002/psc.310010107.

20 Redfield, A.G. (1965). "The theory of relaxation processes." *Adv Magn Reson* 1: 1–32. doi: 10.1016/B978-1-4832-3114-3.50007-6.

21 Goldman, S.A., Bruno, G.V., and Freed, J.H. (1972). "Estimating slow-motional rotational correlation times for nitroxides by electron spin resonance." *J Phys Chem* 76: 1858–1860. doi: 10.1021/j100657a013.

22 Ueda, K., Kimura-Sakiyama, C., Aihara, T. et al. (2011). "Interaction sites of tropomyosin in muscle thin filament as identified by site-directed spin-labeling." *Biophys J* 100 (10): 2432–2439. doi: 10.1016/j.bpj.2011.03.021.

23 Kivelson, D. (1960). "Theory of ESR linewidths of free radicals." *J Chem Phys* 33: 1094–1106. doi: 10.1063/1.1731340.

24 Mason, R.P. and Freed, J.H. (1974). "Estimating microsecond rotational correlation times from lifetime broadening of nitroxide electron spin resonance spectra near the rigid limit." *J Phys Chem* 78: 1321–1323. doi: 10.1021/j100606a015.

25 McConnell, H.M. (1956). "Effect of anisotropic hyperfine inter-actions on paramagnetic relaxation in liquids." *J Chem Phys* 25: 709–711. doi: 10.1063/1.1743033.

26 Mutoh, R., Mino, H., Murakami, R. et al. (2010). "Direct interaction between KaiA and KaiB revealed by a site-directed spin labeling electron spin resonance analysis." *Genes to Cells* 15 (3): 269–280. doi: 10.1111/j.1365-2443.2009.01377.x.

27 Stoll, S. and Schweiger, A. (2006). "EasySpin, a comprehensive software package for spectral simulation and analysis in EPR." *J Magn Reson* 178: 42–55. doi: 10.1016/j.jmr.2005.08.013.

28 Abe, J., Ueki, S., Yamauchi, S. et al. (2018). "Double quantum coherence EPR reveals the structure–function relationships of the cardiac troponin C–troponin I complex regulated by Ca2+ ions and a Phosphomimetic." *Appl Magn Reson* 49 (2018): 893–910. doi: 10.1007/s00723-018-1031-0.

29 Suetake, I., Nakazawa, S., Sato, K. et al. (2021). "Structural dynamics of the chromo-shadow domain and chromodomain of HP1 bound to histone H3K9 methylated peptide, as measured by site-directed spin-labeling EPR spectroscopy." *Biochem Biophys Res Commun* 567: 42–48. doi: 10.1016/j.bbrc.2006.010.

30 Altenbach, C., Greenhalgh, D.A., Khorana, H.G. et al. (1994). "A collision gradient method to determine the immersion depth of nitroxides in lipid bilayers: Application to spin-labeled mutants of bacteriorhodopsin." *Proc Natl Acad Sci U S A* 91 (5): 1667–1671. doi: 10.1073/pnas.91.5.1667.

31 Doebber, M., Bordignon, E., Klare, J.P. et al. (2008). "Salt-driven equilibrium between two conformations in the HAMP domain from Natronomonas pharaonis: The language of signal transfer?" *J Biol Chem* 283 (42): 28691–28701. doi: 10.1074/jbc.M801931200.

32 Lin, Y., Nielsen, R., Murray, D. et al. (1998). "Docking phospholipase A2 on membranes using electrostatic potential-modulated spin relaxation magnetic resonance." *Science* 279 (5358): 1925–1929. doi: 10.1126/science.279.5358.1925.

33 Hubbell, W.L., Mchaourab, H.S., Altenbach, C. et al. (1996). "Watching proteins move using site-directed spin labeling." *Structure* 4(7): 779–783. doi: 10.1016/s0969-2126(96)00085-8.

34 Kageyama, H., Nishiwaki, T., Nakajima, M. et al. (2006). "Cyanobacterial circadian pacemaker: Kai protein complex dynamics in the KaiC phosphorylation cycle in vitro." *Mol Cell* 23: 161–171. doi: 10.1016/j.molcel.2006.05.039.

35 Iwase, R., Imada, K., Hayashi, F. et al. (2005). "Functionally important substructures of circadian clock protein KaiB in a unique tetramer complex." *J Biol Chem* 280: 43141–43149. doi: 10.1074/jbc.M503360200.

36 Pattanayek, R., Wang, J., Mori, T. et al. (2004). "Visualizing a circadian clock protein: Crystal structure of KaiC and functional insights." *Mol Cell* 15: 375–388. doi: 10.1016/j.molcel.2004.07.013.

37 Uzumaki, T., Fujita, M., Nakatsu, T. et al. (2004). "Crystal structure of the C-terminal clock-oscillator domain of the cyanobacterial KaiA protein." *Nat Struct Mol Biol* 11: 623–631. doi: 10.1038/nsmb781.

38 Tseng, R., Goularte, N.F., Chavan, A. et al. (2017). "Structural basis of the day-night transition in a bacterial circadian clock." *Science* 355: 1174–1180. doi: 10.1126/science.aag2516.

39 Snijder, J., Schuller, J., Wiegard, A. et al. (2017). "Structures of the cyanobacterial circadian oscillator frozen in a fully assembled state." *Science* 355 (6330): 1181–1184. doi: 10.1126/science.aag3218.

40 Chang, Y.G., Tseng, R., Kuo, N.W. et al. (2012). "Rhythmic ring-ring stacking drives the circadian oscillator clockwise." *Proc Natl Acad Sci U S A* 109: 16847–16851. doi: 10.1073/pnas.1211508109.

41 Yunoki, Y., Ishii, K., Yagi-Utsumi, M. et al. (2019). "ATP hydrolysis by KaiC promotes its KaiA binding in the cyanobacterial circadian clock system." *Life Sci Alliance* 2 (3): e201900368. doi: 10.26508/lsa.201900368.

42 Yunoki, Y., Matsumoto, A., Morishima, K. et al. (2022). "Overall structure of fully assembled cyanobacterial KaiABC circadian clock complex by an integrated experimental-computational approach." *Commun Biol* 5 (1): 184. doi: 10.1038/s42003-022-03143-z.

43 Ishii, K., Terauchi, S., Murakami, R. et al. (2014). "Site-directed spin labeling-electron spin resonance mapping of the residues of cyanobacterial clock protein KaiA that are affected by KaiA-KaiC interaction." *Genes to Cells* 19 (4): 297–324. doi: 10.1111/gtc.12130.

44 Mutoh, R., Mino, H., Murakami, R. et al. (2011). "Thermodynamically induced conformational changes of the cyanobacterial circadian clock protein KaiB." *Appl Magn Reson* 40: 525–534. doi: 10.1007/s00723-011-0228-2.

45 Chow, G., Chavan, A., Heisler, J. et al. (2020). "Monitoring protein-protein interactions in the cyanobacterial circadian clock in real time via electron paramagnetic resonance spectroscopy." *Biochemistry* 59 (26): 2387–2400. doi: 10.1021/acs.biochem.0c00279.

46 Chiti, F. and Dobson, C.M. (2017). "Protein misfolding, amyloid formation, and human disease: A summary of progress over the last decade." *Annu Rev Biochem* 86: 27–68. doi: 10.1146/annurev-biochem-061516-045115.

47 Sipe, J.D., Benson, M.D., Buxbaum, J.N. et al. (2016). "Amyloid fibril proteins and amyloidosis: Chemical identification and clinical classification International Society of Amyloidosis 2016 Nomenclature Guidelines." *Amyloid* 23 (4): 209–213. doi: 10.1080/13506129.2016.1257986.

48 Török, M., Milton, S., Kayed, R. et al. (2002). "Structural and dynamic features of Alzheimer's Abeta peptide in amyloid fibrils studied by site-directed spin labeling." *J Biol Chem* 277 (43): 40810–40815. doi: 10.1074/jbc.M205659200.

49 Petkova, A.T., Ishii, Y., Balbach, J.J. et al. (2002). "A structural model for Alzheimer's beta -amyloid fibrils based on experimental constraints from solid state NMR." *Proc Natl Acad Sci U S A* 99 (26): 16742–16747. doi: 10.1073/pnas.262663499.

50 Petkova, A.T., Yau, W.M., and Tycko, R. (2006). "Experimental constraints on quaternary structure in Alzheimer's beta-amyloid fibrils." *Biochemistry* 45 (2): 498–512. doi: 10.1021/bi051952q.

51 Cobb, N.J., Sönnichsen, F.D., McHaourab, H. et al. (2007). "Molecular architecture of human prion protein amyloid: A parallel, in-register beta-structure." *Proc Natl Acad Sci U S A* 104 (48): 18946–18951. doi: 10.1073/pnas.0706522104.

52 Der-Sarkissian, A., Jao, C.C., Chen, J. et al. (2003). "Structural organization of alpha-synuclein fibrils studied by site-directed spin labeling." *J Biol Chem* 278 (39): 37530–37535. doi: 10.1074/jbc.M305266200.

53 Margittai, M. and Langen, R. (2004). "Template-assisted filament growth by parallel stacking of tau." *Proc Natl Acad Sci U S A* 101 (28): 10278–10283. doi: 10.1073/pnas.0401911101.

54 Li, E. (2002). "Chromatin modification and epigenetic reprogramming in mammalian development." *Nat Rev Genet* 3 (9): 662–673. doi: 10.1038/nrg887.

55 Bowman, G.D. and Poirier, M.G. (2015). "Post-translational modifications of histones that influence nucleosome dynamics." *Chem Rev* 115 (6): 2274–2295. doi: 10.1021/cr500350x.

56 Leonen, C.J.A., Shimada, M., Weller, C.E. et al. (2021). "Sumoylation of the human histone H4 tail inhibits p300-mediated transcription by RNA polymerase II in cellular extracts." *Elife* 10: e67952. doi: 10.7554/eLife.67952.

57 Nishiyama, A., Yamaguchi, L., and Nakanishi, M. (2016). "Combinations of histone post-translational modifications." *J Biochem* 159 (1): 9–15. doi: 10.1093/jb/mvv113.

58 Rothbart, S.B. and Strahl, B.D. (2014). "Interpreting the language of histone and DNA modifications." *Biochim Biophys Acta* 1839 (8): 627–643. doi: 10.1016/j.bbagrm.2014.03.001.

59 Thålin, C., Aguilera, K., Hall, N.W. et al. (2020). "Quantification of citrullinated histones: Development of an improved assay to reliably quantify nucleosomal H3Cit in human plasma." *J Thromb Haemost* 18 (10): 2732–2743. doi: 10.1111/jth.15003.

60 Noma, K., Allis, C.D., and Grewal, S.I. (2001). "Transitions in distinct histone H3 methylation patterns at the heterochromatin domain boundaries." *Science* 293 (5532): 1150–1155. doi: 10.1126/science.1064150.

61 Barski, A., Cuddapah, S., Cui, K. et al. (2007). "High-resolution profiling of histone methylations in the human genome." *Cell* 129 (4): 823–837. doi: 10.1016/j.cell.2007.05.009.

62 Wang, Z., Zang, C., Rosenfeld, J.A. et al. (2008). "Combinatorial patterns of histone acetylations and methylations in the human genome." *Nat Genet* 40 (7): 897–903. doi: 10.1038/ng.154.

63 Padeken, J., Methot, S.P., and Gasser, S.M. (2022 May 13). "Establishment of H3K9-methylated heterochromatin and its functions in tissue differentiation and maintenance." *Nat Rev Mol Cell Biol* 23 (9): 623–640. doi: 10.1038/s41580-022-00483-w.

64 Nishibuchi, G. and Nakayama, J. (2014). "Biochemical and structural properties of heterochromatin protein 1: Understanding its role in chromatin assembly." *J Biochem* 156 (1): 11–20. doi: 10.1093/jb/mvu032.

65 Dialynas, G.K., Terjung, S., Brown, J.P. et al. (2007). "Plasticity of HP1 proteins in mammalian cells." *J Cell Sci* 120 (Pt 19): 3415–34124. doi: 10.1242/jcs.012914.

66 Mishima, Y., Jayasinghe, C.D., Lu, K. et al. (2015). "Nucleosome compaction facilitates HP1γ binding to methylated H3K9." *Nucleic Acids Res* 43 (21): 10200–10212. doi: 10.1093/nar/gkv841.

67 Mishima, Y., Watanabe, M., Kawakami, T. et al. (2013). "Hinge and chromoshadow of HP1α participate in recognition of K9 methylated histone H3 in nucleosomes." *J Mol Biol* 425 (1): 54–70. doi: 10.1016/j.jmb.2012.10.018.

68 Watanabe, S., Mishima, Y., Shimizu, M. et al. (2018). "Interactions of HP1 bound to H3K9me3 dinucleosome by molecular simulations and biochemical assays." *Biophys J* 114 (10): 2336–2351. doi: 10.1016/j.bpj.2018.03.025.

69 Nielsen, P.R., Nietlispach, D., Mott, H.R. et al. (2002). "Structure of the HP1 chromodomain bound to histone H3 methylated at lysine 9." *Nature* 416 (6876): 103–107. doi: 10.1038/nature722.

70 Canzio, D., Liao, M., Naber, N. et al. (2013). "A conformational switch in HP1 releases auto-inhibition to drive heterochromatin assembly." *Nature* 496 (7445): 377–381. doi: 10.1038/nature12032.

71 Suetake, I., Sato, K., Sugishita, T. et al. (2022). "Dynamics of the HP1 Hinge Region with DNA Measured by Site-directed Spin Labeling-EPR Spectroscopy." *Appl Magn Reson* (in press).

72 Fichou, Y., Lin, Y., Rauch, J.N. et al. (2018). "Cofactors are essential constituents of stable and seeding-active tau fibrils." *Proc Natl Acad Sci U S A* 115 (52): 13234–13239. doi: 10.1073/pnas.1810058115.

73 Kumari, P., Ghosh, D., Vanas, A. et al. (2021). "Structural insights into α-synuclein monomer-fibril interactions." *Proc Natl Acad Sci U S A* 118 (10): e2012171118. doi: 10.1073/pnas.2012171118.

74 Lin, Y., McCarty, J., Rauch, J.N. et al. (2019). "Narrow equilibrium window for complex coacervation of tau and RNA under cellular conditions." *eLife* 8: e42571. doi: 10.7554/eLife.42571.

75 Emmanouilidis, L., Esteban-Hofer, L., Damberger, F.F. et al. (2021). "NMR and EPR reveal a compaction of the RNA-binding protein FUS upon droplet formation." *Nat Chem Biol* 17 (5): 608–614. doi: 10.1038/s41589-021-00752-3.

76 Larson, A.G., Elnatan, D., Keenen, M.M. et al. (2017). "Liquid droplet formation by HP1α suggests a role for phase separation in heterochromatin." *Nature* 547 (7662): 236–240. doi: 10.1038/nature22822.

77 Strom, A.R., Emelyanov, A.V., Mir, M. et al. (2017). "Affiliations expand. Phase separation drives heterochromatin domain formation." *Nature* 547 (7662): 241–245. doi: 10.1038/nature22989.

78 Keenen, M.M., Brown, D., Brennan, L.D. et al. (2021). "HP1 proteins compact DNA into mechanically and positionally stable phase separated domains." *Elife* 10: e64563. doi: 10.7554/eLife.64563.

79 Wang, L., Gao, Y., Zheng, X. et al. (2019). "Histone modifications regulate chromatin compartmentalization by contributing to a phase separation mechanism." *Mol Cell* 76 (4): 646–659. doi: 10.1016/j.molcel.2019.08.019.

80 Erdel, F., Rademacher, A., Vlijm, R. et al. (2020). "Mouse heterochromatin adopts digital compaction states without showing hallmarks of HP1-driven liquid-liquid phase separation." *Mol Cell* 78 (2): 236–249. doi: 10.1016/j.molcel.2020.02.005.

2

Introduction to Incoherent Neutron Scattering

A Powerful Technique to Investigate the Dynamics of Bio-macromolecules

Tatsuhito Matsuo[1,2,3,*] *and Judith Peters*[1,2,4]

[1] *Univ. Grenoble Alpes, CNRS, LiPhy, Grenoble, France*
[2] *Institut Laue-Langevin, Grenoble Cedex 9, France*
[3] *Institute for Quantum Life Science, National Institutes for Quantum Science and Technology, Tokai, Ibaraki, Japan*
[4] *Institut Universitaire de France, Paris, France*
* *Corresponding author*

2.1 Introduction

In living organisms, bio-macromolecules, e.g. proteins, lipids, nucleic acids, express their specific functions in aqueous environments as in their native environment, the cells. The molecules are constantly varying under the influence of thermal energy and make use of the conformational fluctuations to change their structure, which is a prerequisite for expressing functions such as enzymatic activity. Focusing on proteins, it is well known that various motions contained in the molecules obey a hierarchy [1], as shown in Figure 2.1. The global diffusion represents the center-of-mass translational and rotational motions of the whole molecule. This motion plays an important role in biochemical reactions such as protein–protein associations, where the rate-limiting step is the diffusion of the protein. In the case where the protein molecules consist of several domains, there exist domain movements, which can be seen, for example, in molecular motors such as skeletal muscle myosin [2]. More localized segmental and backbone motions contribute to the function of intrinsically disordered proteins or flapping motions of an enzyme to capture ligands as observed for the HIV protease [3]. The side chains of proteins directly interact with other chemical groups of their

Analytical Techniques for the Elucidation of Protein Function, First Edition. Edited by Isao Suetake, Rohit K. Sharma, and Hironobu Hojo.

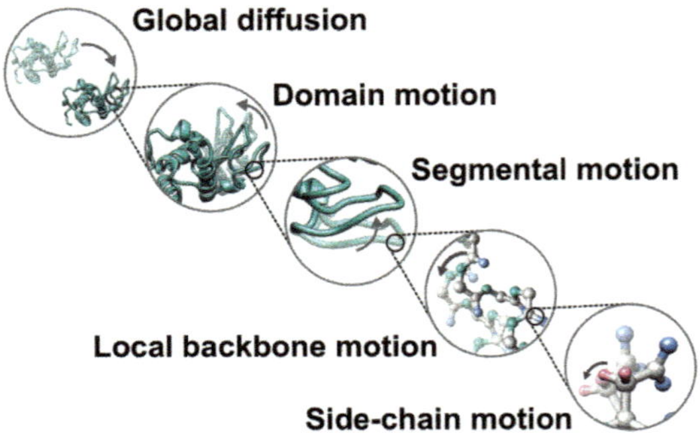

Figure 2.1 An illustration of the hierarchy of different types of motions contained in protein molecules. In the lower two drawings, carbon, nitrogen, oxygen, and hydrogen atoms are depicted in gray, cornflower blue, light sea green, and hot pink, respectively. As a protein, a lysozyme structure (PDB ID: 1jpo) is used in this figure. The molecular images were produced using UCSF Chimera (which was also used for other figures in this chapter) [70].

binding partner or ligands or with water molecules, and thus their mobility plays a role in binding processes or enzymatic activity. For instance, inhibitor binding to human acetylcholinesterase changes side-chain motions [4]. In addition to the motions described in Figure 2.1, there are vibrations of atoms and bond fluctuations contributing to the chemical bond strength and interactions between chemical groups. The localized atomic motions explained here can be simulated by molecular dynamics (MD) simulation, which has the time- and space scales compatible with incoherent neutron scattering, the main theme of this chapter. Combined use of MD simulations and neutron scattering provides insights into the relevant atomic motions contained in the investigated samples [5].

In general, it is thought that specific types of motions of a given protein in the entire dynamical hierarchy are related to its function (e.g. a majority of research on molecular motors such as myosin focuses on the domain movements). However, it has been shown that, for instance, pico- to nanosecond timescale atomic fluctuations facilitate the large-scale slower motions necessary for catalytic function of adenylate kinase [1, 6, 7], suggesting that the motions in different tiers of the dynamical hierarchy are closely related with each other even in the same protein. Therefore, it is essential to gain deep knowledge of molecular dynamics at different temporal and spatial scales and then integrate all the information to profoundly understand the mechanism of the biological functions of a given bio-macromolecule.

One of the powerful experimental techniques to observe dynamical behavior of bio-macromolecules is incoherent neutron scattering (iNS), which is described in this chapter. In brief, iNS provides dynamical information at the angstrom length scale and at the pico- to nanosecond timescales (see Figure 2.1) depending on the specification of neutron spectrometers. Furthermore, this technique has the advantage that the samples can be in any state, e.g. powder, solution, crystal, or even cell pellets. The pico- to nanosecond dynamics serve as a driving force to realize large-scale slower motions taking place at micro- and millisecond timescales [6–8]. Thus, iNS is an important technique to characterize the molecular dynamics of proteins, and it has been used to relate dynamical behavior of proteins to their biological functions [5, 9].

iNS measurements are carried out using spectrometers installed at large-scale facilities such as a research reactor, as the Institut Laue Langevin (ILL) in Grenoble, France, or a spallation source, as J-PARC in Tokai, Japan, or the SNS in Oak Ridge, USA (a list of large-scale neutron facilities can be found on the website https://www.iucr.org/resources/commissions/neutron-scattering/where-neutrons). Compared with laboratory-scale experiments, there are less opportunities to utilize this technique due to the limited availability of such facilities. The situation is more or less the same for X-ray or neutron crystallography despite the availability of laboratory-scale instruments for X-ray crystallography. However, unlike iNS, X-ray and neutron crystallography have become widely used techniques and most importantly not only physicists but also molecular biologists and biochemists are increasingly utilizing these techniques for their own research. This is partly due to the development of a series of excellent user-friendly software to analyze diffraction data, which does not require deep knowledge of physics. On the other hand, there is no systematic software to analyze the iNS data and extract dynamical information automatically because analysis of the iNS data depends on the nature of the samples and the physical quantities that one wants to observe. Moreover, the analysis and interpretation of the iNS data requires a certain amount of knowledge on the atomic and molecular motions, which falls into the scope of different fields of physics such as nonequilibrium statistical physics and quantum physics. Considering these points, a brief and concise overview of iNS is essential in order to disseminate this technique with the aim of establishing the dynamics-function relationship in a wider variety of bio-macromolecules studied in biochemistry, molecular biology, and physiology.

In this chapter, a basic theory of iNS is first described, corresponding to a minimum requirement to understand the principles of this technique and analyze the experimental data. Then, a series of dynamical information that can be extracted from the iNS spectra are described. Finally, some examples of biological applications of iNS are presented, putting emphasis on the role of sub-nanosecond molecular dynamics in biological functions of a given system. Note that these examples do not cover comprehensively this research field, and various research groups all over the world are working on the dynamics of bio-macromolecules focusing on different physicochemical aspects of each target. Readers who are interested in more detailed principles, theories, and applications are referred to the literature [10–14].

2.2 Basic Theory and Dynamical Information Obtained from iNS

2.2.1 Basic Principle of iNS Experiments

A typical experimental configuration of iNS experiments is shown in Figure 2.2a. Incident neutron beams are generated by a research reactor or a spallation source. In a research reactor, neutrons are produced through nuclear fission of uranium (^{235}U) while in a spallation source, neutrons are produced by a spallation reaction, where a target material such as mercury is bombarded with high-energy protons, which destroy the nuclei of atoms in the target, thereby emitting neutrons. Since neutron production requires these reactions, laboratory-scale neutron scattering instruments do not exist at the moment unlike the X-ray scattering instruments.

The energy of the produced neutrons is too high to be used for scattering measurements. Therefore, the energy of the neutrons needs to be decreased by collisions with materials such as light water (H_2O) or liquid hydrogen (a container filled with these materials is called a moderator). In scattering experiments, neutrons with wavelength of 1–10 Å (0.8–82 meV) are typically used. The neutron energy and its wavelength are linked by the following relation:

$$E = \frac{81.81}{\lambda^2}, \tag{2.1}$$

where E (meV) is the neutron energy and λ (Å) its wavelength. Neutrons with wavelengths of 1–3 Å and > 3 Å are called thermal and cold neutrons, respectively, although the boundary between them depends on the literature.

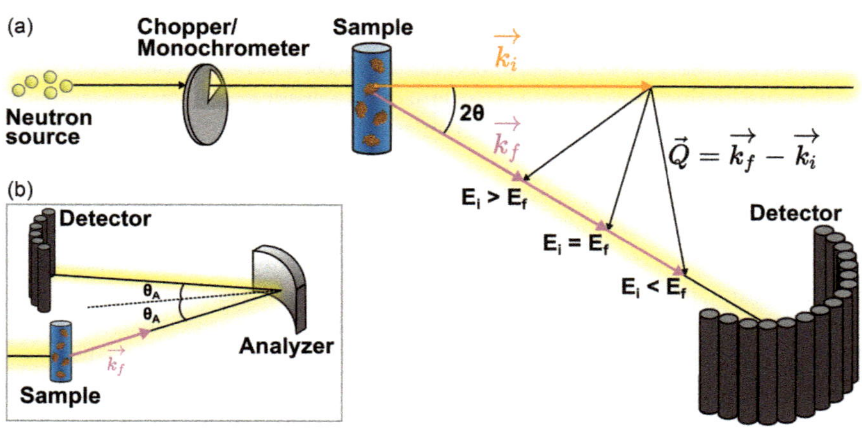

Figure 2.2 (a): A schematic of an incoherent neutron scattering experiment for a time-of-flight configuration. For the definition of the symbols in the figure, readers are referred to the main text. (b): A schematic for a backscattering configuration.

The moderated neutron beam still contains neutrons with different energies. In order to generate an incident neutron beam suitable for scattering measurements, the neutrons are subjected to a chopper or a monochromator (Figure 2.2a), which narrows down the energy range of the neutrons or singles out neutrons with a specific energy, respectively. The resultant neutron beam irradiates the sample to be investigated, which can be in any state: crystals, solutions, powders, or even cell pellets. The required amount of biomolecules in the samples depends on spectrometers and is typically in the range of 20–200 mg. When neutrons enter the inside of the sample, they interact with the atomic nuclei. Some neutrons penetrate or are absorbed by the sample, while others are scattered in the direction forming an angle 2θ with respect to the incident beam. Here, we define the wave vector ($\vec{k}$) whose modulus is

$$k = \left|\vec{k}\right| = \frac{2\pi}{\lambda}, \tag{2.2}$$

where k is the wavenumber. The wave vectors of the incident neutron and of the scattered neutron are defined as $\overrightarrow{k_i}$ and $\overrightarrow{k_f}$, respectively. The momentum vector of the neutron is described as

$$\vec{p} = \hbar\vec{k}, \tag{2.3}$$

where $\hbar$ is the reduced Planck's constant or Dirac's constant ($\hbar = 1.055 \times 10^{-34}\,\mathrm{J \cdot s}$). Then, the momentum transfer of the neutron, which is the change in momentum caused by the interaction between the incident neutron and a nucleus in the sample, is described by

$$\overrightarrow{p_f} - \overrightarrow{p_i} = \hbar\left(\overrightarrow{k_f} - \overrightarrow{k_i}\right) = \hbar\vec{Q}, \tag{2.4}$$

where $\overrightarrow{p_i}$ and $\overrightarrow{p_f}$ are the momentum vectors of the incident and the scattered neutron, respectively, and $\vec{Q} = \overrightarrow{k_f} - \overrightarrow{k_i}$. Thus, $\vec{Q}$ denotes the momentum transfer of the neutron as multiple of $\hbar$ and is an important physical quantity in scattering measurements.

As a result of the interaction between a neutron and an atomic nucleus in the sample, the scattered neutron either maintains its energy or exchanges energy with the atom. The energy of a neutron as introduced in Eq. 2.1 can also be expressed in the form of

$$E = \frac{1}{2}m\left|\vec{v}\right|^2 = \frac{\hbar^2\left|\vec{k}\right|^2}{2m} = \hbar\omega, \tag{2.5}$$

where $\vec{v}$ is the velocity vector of the neutron, m is the neutron mass (1.675×10^{-27} kg), and ω is the angular frequency. In the case where there is no energy exchange, i.e. $E_i = E_f$, where E_i, E_f are the energy of the incident and the scattered neutron, respectively, the scattering process is called elastic scattering. In this case, as easily calculated, the modulus of $\vec{Q}$ is

$$\left|\vec{Q}\right| = \frac{4\pi \sin\theta}{\lambda} = 2\left|\vec{k}\right| \sin\theta. \tag{2.6}$$

On the other hand, in the case where there is an energy exchange between a neutron and an atom in the sample, the scattered neutron loses energy ($E_i > E_f$) or gains energy ($E_i < E_f$). In this case, the scattering process is called inelastic scattering. In particular, when $\Delta E = |E_f - E_i|$ is relatively small, the scattering process is called quasi-elastic scattering. In this case, the modulus $\vec{Q}$ is calculated from the law of cosines as

$$\left|\vec{Q}\right| = \sqrt{\left|\vec{k_i}\right|^2 + \left|\vec{k_f}\right|^2 - 2\left|\vec{k_i}\right|\left|\vec{k_f}\right| \cos 2\theta}\ . \tag{2.7}$$

It is easily found that when $\left|\vec{k_i}\right| = \left|\vec{k_f}\right|$ (i.e. elastic scattering), Eq. 2.7 reduces to Eq. 2.6.

After the interaction with atoms in the sample, the scattered neutrons are counted in neutron detectors. A neutron is hard to detect because it interacts only weakly with matter and it does not have any electric charge. In many spectrometers, scattered neutrons are detected using ^{3}He scintillation detectors making use of the following nuclear reaction:

$$n + {}^3He \rightarrow {}^3H + p + 764\ keV, \tag{2.8}$$

where n and p are a neutron and a proton (H^+), respectively. Following this reaction, neutrons can be recorded through detection of protons, i.e. charged particles, which are easily measured. It is worth noting that although the neutron does not have any charge, it has a magnetic moment. According to the quark model, the neutron consists of one up quark and two down quarks [15]. The up and down quarks carry the charge of +(2/3)e and –(1/3)e, respectively, giving the neutron a net charge of zero. On the other hand, the magnetic moment of the neutron is represented as the sum of the magnetic moments of these quarks. Therefore, despite being a non-charged particle, the neutron has a magnetic moment.

From the above explanation, it follows that the momentum transfer ($\vec{Q}$) and the energy transfer ($\Delta E = \Delta\hbar\omega$) characterize the interaction between neutrons and atoms in the sample. Because the number of scattered neutrons detected at specific $\vec{Q}$ and ΔE values depends on the nature of atomic motions in the sample, analysis of the scattering spectra (or the dynamic scattering function) $S(\vec{Q}, \Delta E)$ or $S(\vec{Q}, \omega)$ provides the dynamical information of the samples. According to Heisenberg's uncertainty principle, the energy transfer and the corresponding timescale (Δt) are related to each other by

$$\Delta E \cdot \Delta t \geq \frac{\hbar}{2}. \tag{2.9}$$

This can be written in a more practical form:

$$\Delta t \geq \frac{658.55}{(2\Delta E)}, \tag{2.10}$$

where ΔE is in μeV and Δt is in picoseconds. $2\Delta E$ corresponds to the full width at half maximum (FWHM) of $S(\vec{Q}, \omega)$ as discussed later. The $2\Delta E$ values specific to each neutron spectrometer are called the energy resolution, which defines the timescale (the time window) that is accessible by the spectrometer. The resolution function depends on the instrumental setup, and thus the combined use of several spectrometers permits to study the dynamics of the sample in wider timescales. In Figure 2.2, two types of instrumental configurations are shown. In Figure 2.2a, the scattered neutron goes directly into the detector, and thus its energy is calculated from the distance between the sample and the detector and the time spent on traveling the distance. This type of instrument is called time-of-flight (TOF) spectrometer. On the other hand, in Figure 2.2b, an additional component, e.g. Si crystals, called analyzers, is installed between the sample and the detector. The energy resolution in this case is expressed by [11]

$$\frac{\Delta E}{E} = 2\cot(90° - \theta_A)\Delta(90° - \theta_A) + \frac{2\Delta d}{d}, \tag{2.11}$$

where E is the energy of the incident neutron, θ_A is the incident (or the reflected) angle of the neutron on the analyzer surface, and d is the lattice spacing of the analyzer crystal. The second term depends on the quality of the analyzer crystal. Since $\cot(90° - \theta_A)$ approaches zero as θ_A approaches 0°, the lowest energy resolution is achieved when θ_A is close to 0° (i.e. the neutron is scattered in a backward direction). This type of instrument is called backscattering spectrometer (BS), and its energy resolution is generally smaller than that of TOF instruments. Conventionally, when the energy resolution (ΔE) is small, it is called "high" energy resolution and vice versa.

2.2.2 Incoherent Scattering Function

During the scattering process, neutrons interact with the nuclei of atoms in the sample. The scattering capacity of the atoms is called the scattering cross-section, which is usually denoted by σ. The most important physical quantity in iNS measurements is the partial differential cross-section, which is defined by [12]

$$\frac{d^2\sigma}{d\Omega dE} = \frac{N_{scat}}{\Phi d\Omega dE} = C_{instrument}\, S(\vec{Q}, \omega), \tag{2.12}$$

where N_{scat} is the number of neutrons scattered per second into a solid angle $d\Omega$ with the energy between E and E + dE, and Φ is the flux of the incident neutron beam with the unit of $cm^{-2}s^{-1}$. Since the detectors installed in neutron spectrometers count the number of neutrons scattered in their direction, the partial differential cross-section is proportional to the experimentally obtained scattering spectra $S(\vec{Q}, \omega)$. The factor $C_{instrument}$, which depends on the instrument and the experimental setup, is necessary to represent $S(\vec{Q}, \omega)$ on the absolute scale. In the following, we assume $C_{instrument} = \frac{1}{4\pi}\frac{k_f}{k_i}$ for simplicity (although measurements on the absolute scale are not common in iNS, the interested reader can find a method to determine this factor as often used in small-angle neutron scattering [16]).

Now, let us consider a sample consisting of N identical atoms. For this system, the time-dependent pair-correlation function and the self time-dependent pair-correlation function are defined as [12]

$$G(\vec{r},t) = \frac{1}{N}\sum_{i,j=1}^{N}\left\langle \delta\left\{\vec{r} - \vec{R_i}(t) + \vec{R_j}(0)\right\}\right\rangle, \tag{2.13}$$

$$G_s(\vec{r},t) = \frac{1}{N}\sum_{i=1}^{N}\left\langle \delta\left\{\vec{r} - \vec{R_i}(t) + \vec{R_i}(0)\right\}\right\rangle, \tag{2.14}$$

where $\vec{r}$ is the position vector, and $\vec{R_i}(t)$ is the position vector of an i-th atom at time t. $\langle\,\rangle$ denotes the thermal average, where $\delta\left\{\vec{r} - \vec{R_i}(t) + \vec{R_j}(0)\right\}$ is averaged over all the possible configurations. Note that in the scattering theory, these physical quantities are treated in the framework of quantum physics and thus are represented as operators. However, in this chapter, we consider only the case of the classical limit for simplicity, where classical mechanics applies. This approximation is often used in neutron scattering theory [5, 11, 12] and is valid as long as the exchanged energies are small compared to the thermal energy ½ k_B T. As seen in Eq. 2.13, $G(\vec{r},t)$ describes the probability that a particle resides at position $\vec{r}$ at time t, given that another particle or an identical particle was situated at the origin at time $t = 0$. Thus, this function includes the correlation between identical or different atoms. On the other hand, $G_s(\vec{r},t)$ describes the probability that a particle resides at position $\vec{r}$ at time t, given the same particle was at the origin at time $t = 0$. This function thus contains only the correlation between identical atoms at different times.

The partial differential cross-section of the above sample is expressed as [11, 12]

$$\begin{aligned}\left(\frac{d^2\sigma}{d\Omega dE}\right)_{sample} &= \left(\frac{d^2\sigma}{d\Omega dE}\right)_{coh} + \left(\frac{d^2\sigma}{d\Omega dE}\right)_{inc} \\ &= \frac{\sigma_{coh}}{4\pi}\frac{k_f}{k_i}S_{coh}\left(\overline{Q},\omega\right) + \frac{\sigma_{inc}}{4\pi}\frac{k_f}{k_i}S_{inc}\left(\overline{Q},\omega\right),\end{aligned} \tag{2.15}$$

$$S_{coh}\left(\vec{Q},\omega\right) = \frac{1}{2\pi\hbar}\iint G\left(\vec{r},t\right)\exp\left\{i\left(\vec{Q}\vec{r} - \omega t\right)\right\}d\vec{r}dt, \tag{2.16}$$

$$S_{coh}\left(\vec{Q},\omega\right) = \frac{1}{2\pi\hbar}\iint G\left(\vec{r},t\right)\exp\left\{i\left(\vec{Q}\vec{r} - \omega t\right)\right\}d\vec{r}dt, \tag{2.17}$$

where subscripts coh and inc denote the coherent and incoherent scattering, respectively. $S_{coh}\left(\vec{Q},\omega\right)$ and $S_{coh}\left(\vec{Q},\omega\right)$ are called the coherent and the incoherent scattering function, respectively. As seen in Eqs. 2.16 and 2.17, the scattering functions are the Fourier transforms in time and space of the pair-correlation functions defined above. Therefore, $S_{coh}\left(\vec{Q},\omega\right)$ contains the information on the distances between different atoms in the sample at different times, meaning that $S_{coh}\left(\vec{Q},\omega\right)$ includes the information on the structure of the sample. On the other hand, $S_{coh}\left(\vec{Q},\omega\right)$ contains the information on the positions of the same atom at different times, meaning that $S_{inc}\left(\vec{Q},\omega\right)$ permits to monitor the movement of identical atoms in the sample. Thus, $S_{inc}\left(\vec{Q},\omega\right)$ is the most important function to investigate molecular dynamics.

In Eq. 2.15, σ_{coh} and σ_{inc} are the coherent and the incoherent scattering cross-section of the sample, respectively. The σ_{coh} and σ_{inc} values of atoms typically found in bio-macromolecules and buffer solutions are shown in Figure 2.3.

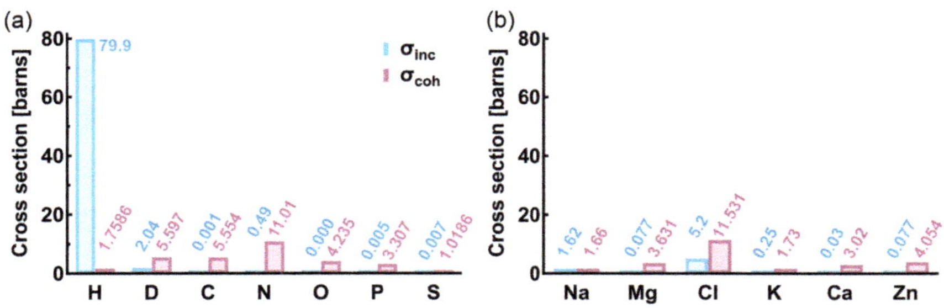

Figure 2.3 Incoherent and coherent scattering cross-sections of different atoms relevant to bio-macromolecules. (a): Scattering cross-sections of the atoms constituting the protein, the lipid, and the nucleic acid and of an isotope of hydrogen atoms, i.e. deuterium (D). (b): Those of the atoms often utilized as ligands and solutes of the buffer solutions. Note that 1 barn = 10^{-24} cm^2. The values of the cross-section in this figure were adapted from [11]. For reference, a comprehensive list of the scattering cross-sections is found in [17].

Those of other atoms are tabulated in the literature [11, 17]. From Figure 2.3, it is seen that σ_{inc} of the hydrogen atom is much larger than that of its isotope, i.e. deuterium (D), and any other atom, showing that the incoherent scattering signal from hydrogen atoms dominates the experimentally obtained spectra in iNS. In the case of solution samples, the use of D_2O-based buffers as solvent will thus permit to obtain the scattering signal from nonexchangeable hydrogen atoms in bio-macromolecules at a high signal-to-noise ratio. Here, exchangeable (or labile) hydrogen atoms refer to those which bind to O or N and thus tend to be replaced by D contained in the solvent. In the following, the term "hydrogen atoms" means nonexchangeable hydrogen atoms unless otherwise stated. Since hydrogen atoms are quasi-uniformly distributed throughout the proteins and lipids as shown in Figure 2.4, the atomic motions averaged over the whole molecule are observed using iNS. In the case of nucleic acids, ~75% of all the hydrogen atoms are in the sugar riboses and the remaining ~25% in the bases. Thus, the motions of hydrogen atoms in nucleic acids reflect those of the main chain.

As evident from Eq. 2.15, the partial differential cross-section of the sample contains both the coherent and incoherent signals. Although the incoherent scattering cross-section of hydrogen atoms is much larger than the coherent scattering cross-section of other atoms as shown in Figure 2.3, it is important to estimate the fraction of coherent and incoherent scattering signals in the partial differential cross-section of the sample. For this purpose, we focus on $S(\vec{Q})$, which is obtained by integrating $S(\vec{Q},\omega)$ over the energy transfer. From Eqs. 2.12 and 2.15, $S(\vec{Q},\omega)$ reads

$$S(\vec{Q},\omega)=\sigma_{coh}S_{coh}(\vec{Q},\omega)+\sigma_{inc}S_{inc}(\vec{Q},\omega). \tag{2.18}$$

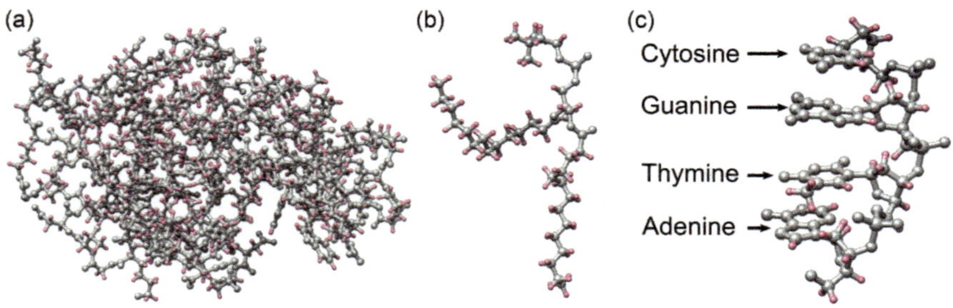

Figure 2.4 Spatial distribution of nonexchangeable hydrogen atoms in (a) a protein, (b) a lipid, and (c) a nucleic acid (DNA). The PDB ID of the protein structure is 1JPO. The lipid structure is of DMPC, which was obtained from https://people.ucalgary.ca/~tieleman/download.html, http://www.scfbio-iitd.res.in/software/drugdesign/bdna.jsp#. The DNA structure was generated using the online server http://www.scfbio-iitd.res.in/software/drugdesign/bdna.jsp#. Hydrogen atoms are shown in hot pink while other atoms are in gray in all the panels.

Thus, $S(\vec{Q})$ reads

$$S(\vec{Q}) = \sigma_{coh} S_{coh}(\vec{Q}) + \sigma_{inc} S_{inc}(\vec{Q}). \tag{2.19}$$

The integration of Eq. 2.19 over $\vec{Q}$ gives the total scattering cross-section $\sigma_{tot}(= \sigma_{coh} + \sigma_{inc})$ of the sample. Here, we consider protein solution samples, where protein molecules orient toward random directions. In this case, due to the spatial averaging, the momentum transfer is expressed by its modulus Q. In order to see the fraction of incoherent scattering in $S(Q)$, the two terms in Eq. 2.19, i.e. $\sigma_{coh} S_{coh}(Q)$ and $\sigma_{inc} S_{inc}(Q)$, need to be compared. In Figure 2.5, a rough comparison was made using simulated $S_{coh}(Q)$ curves, which were calculated by CRYSON [18]. Here, the solvent contribution is subtracted during the calculation of $S_{coh}(Q)$ curves in CRYSON. Regarding $S_{inc}(Q)$, by integrating Eq. 2.17 over ω and substituting Eq. 2.14,

$$\begin{aligned}
S_{inc}(\vec{Q}) &= \int S_{inc}(\vec{Q}, \omega) d\omega \\
&= \frac{1}{2\pi\hbar} \iint G_s(\vec{r}, t) \left(\int \exp(-i\omega t) d\omega \right) \exp(i\vec{Q}\vec{r}) d\vec{r} dt \\
&= \frac{1}{2\pi\hbar} \iint G_s(\vec{r}, t) \delta(t) \exp(i\vec{Q}\vec{r}) d\vec{r} dt \\
&= \frac{1}{2\pi\hbar} \int G_s(\vec{r}, 0) \exp(i\vec{Q}\vec{r}) d\vec{r} \\
&= \frac{1}{2\pi\hbar} \int \frac{1}{N} \sum_{i=1}^{N} \langle \delta(\vec{r}) \rangle \exp(i\vec{Q}\vec{r}) d\vec{r} \\
&= \frac{1}{2\pi\hbar} \frac{1}{N} N = \frac{1}{2\pi\hbar} = \mathit{Constant}
\end{aligned} \tag{2.20}$$

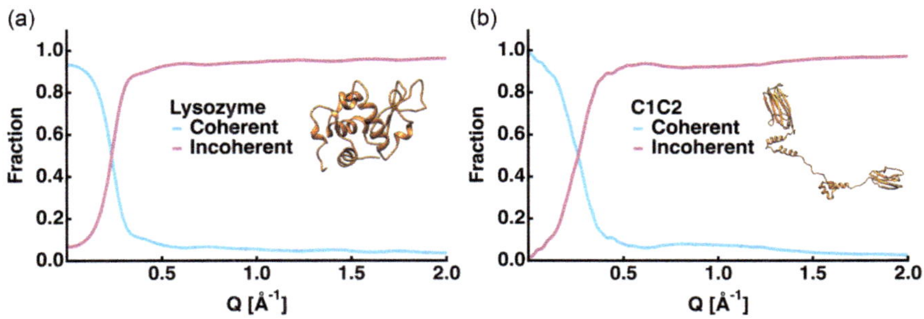

Figure 2.5 Simulation-based estimation of the fractions of coherent and incoherent scattering in the ω-integrated scattering function S(Q) (see Eq. 2.19 in the main text). As a globular protein and an elongated protein, lysozyme (a) and a structural model of C1C2 (b) were used. Coherent and incoherent fractions are denoted in cyan and magenta, respectively. The PDB ID of the lysozyme structure is 1JPO, and the C1C2 model is a randomly created structure [19].

$S_{inc}(Q)$ was thus assumed to be constant. In Eq. 2.20, the following well-known properties of the Dirac's delta function are used:

$$\int \exp(i\omega x)d\omega = \delta(x), \tag{2.21}$$

$$\int f(x)\delta(x-a)dx = f(a), \tag{2.22}$$

where $f(x)$ is an arbitrary function of variable x and a is a constant. Here, models of a globular protein (lysozyme) and an elongated protein (C1C2 [19]) were employed to see the possible effects of protein size. The maximum dimensions of these proteins are ~47 Å and ~150 Å, respectively. It is found that regardless of protein size, the coherent scattering curve drastically decreases when Q increases and is almost constant in the range of $Q > 0.4$ Å^{-1}. As a result, the incoherent scattering contributes by 90–95% at $Q > 0.4$ Å^{-1}. Using polarized neutron scattering, it has experimentally been shown that in the case of myoglobin, the incoherent contribution was ~90% [20]. Therefore, to avoid the "contamination" by coherent scattering and focus on incoherent scattering, it would be safe to analyze the measured $S(Q,\omega)$ data at $Q > 0.4$ Å^{-1}. If results from a polarization neutron scattering experiment are not available, it might be useful to consider utilizing the molecular structures, which may be retrieved from Protein Data Bank (https://www.rcsb.org), in order to calculate the corresponding $S_{coh}(Q)$, i.e. the small-angle neutron scattering (SANS) curve. In SANS, incoherent scattering is treated as a "background" unlike iNS. Regarding the basics of SANS technique, readers might be referred to the literature [21].

In actual iNS experiments, $S(Q,\omega)$ is obtained from the raw data after data reduction by specific software, which is distributed from various neutron facilities. For example, the following software is available:

LAMP [22] (https://www.ill.eu/users/support-labs-infrastructure/software-scientific-tools/lamp)
DAVE [23] (https://www.ncnr.nist.gov/dave/download.html)
UTSUSEMI [24] (https://mlfinfo.jp/groups/comp/en/utsusemi.html)
Mantid [25] (https://www.mantidproject.org/Main_Page.html)

LAMP, DAVE, and UTSUSEMI were developed by the ILL in France, National Institute of Standards and Technology in the United States, and J-PARC in Japan, respectively. Mantid is a project as a collaboration between many neutron facilities around the world. Based on the reduced data, various parameters characterizing the dynamics of the sample (see Section 2.2.3) can be extracted by a fitting approach of $S(Q,\omega)$. Although some fitting programs are already included in the above software, it is often convenient to produce proprietary fitting programs using other software such as IGOR PRO (WaveMetrics, Lake Oswego, Oregon, United States).

2.2.3 Dynamical Information Obtained by iNS

iNS generally includes three techniques called elastic incoherent neutron scattering (EINS), quasi-elastic neutron scattering (QENS), and inelastic neutron scattering (INS). As is clear from their names, EINS measures neutrons scattered elastically, i.e. without changing their energy during the scattering process (see also Figure 2.2), and QENS measures neutrons scattered with slight changes in energy during the scattering process in addition to elastically scattered neutrons. Since the quasi-elastic scattering is inelastic scattering with only a slight energy change in the scattered neutrons, QENS is the limiting case of INS.

A schematic illustration of iNS spectra at an arbitrary Q value is shown in Figure 2.6. The iNS spectra consist of the elastic component, the quasi-elastic component, and the inelastic component. Various motions in bio-macromolecules are classified into these components (or the background) of the spectra according to the timescale of the motions. The time scale or the energy resolution that is accessible by the spectrometer employed is often determined by the spectra of vanadium, which is a totally incoherent and elastic scatterer. The FWHM of the resultant elastic component ($2\Delta E_{elastic}$) is the energy resolution of the spectrometer, which can be converted to the corresponding timescale using Eq. 2.10. For example, if $2\Delta E_{elastic}$ of the vanadium spectrum is 10 μeV, the corresponding timescale is $\Delta t \sim 66$ ps. Since the motions that have larger $2\Delta E$ values than the energy resolution (i.e. motions with shorter relaxation times) are also resolved, the accessible time window is defined as $\Delta t < 66$ ps in this case. On the other hand, in the case of slower motions with relaxation times larger than 66 ps, the corresponding energy width $2\Delta E$ is smaller than that of the resolution function. Thus, for motions

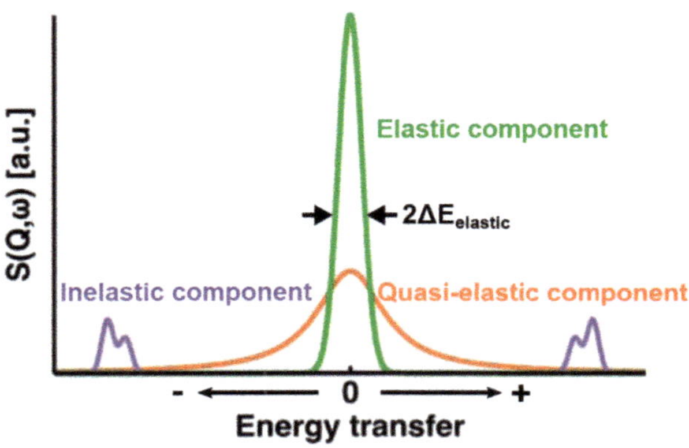

Figure 2.6 A schematic illustration of incoherent neutron scattering spectra $S(Q, \omega)$ at an arbitrary Q value. The full width at half maximum (FWHM) of the elastic component is denoted by $2\Delta E_{elastic}$. The corresponding timescale is calculated from Eq. 2.10 in the main text.

with $\Delta t >> 66$ ps, the corresponding elastic or quasi-elastic component cannot be distinguished from the spectra of the resolution function, meaning that these motions are not resolved. However, it has been shown that iNS can resolve the motions with their energy widths down to ~10% of the energy resolution [26, 27]. In the following, among the three techniques of iNS, we focus on EINS and QENS, which are more often used for dynamics studies on bio-macromolecules, and then we summarize dynamical information obtained from EINS and QENS.

2.2.3.1 Elastic Incoherent Neutron Scattering (EINS)

The EINS scattering intensity (or EINS curve) is a function of Q with zero energy transfer, and is thus denoted by $S(Q,\omega)|_{\omega\approx 0} = S(Q,0)$ in an ideal case. However, in practice, because of the finite energy resolution (ΔE) of the spectrometer employed, the scattering intensity is integrated over the energy width corresponding to the energy resolution. Therefore, the scattering intensity is denoted by $S(Q,\pm\Delta E)$ or $S(Q,\pm\Delta\omega)$. At smaller-Q values, $S(Q,\pm\Delta E)$ can be approximated by a Gaussian function [28] assuming only harmonic motions of the atoms around their equilibrium positions,

$$S(Q,\pm\Delta E) = S(0,\pm\Delta E)\exp\left(-\frac{Q^2\langle u^2\rangle}{3}\right) \tag{2.23}$$

where $S(0,\pm\Delta E)$ is the scattering intensity at $Q = 0$ [Å^{-1}] and $\langle u^2\rangle$ is the mean square displacement of atoms. Whereas this approximation is rigorously valid under the condition $Q^2\langle u^2\rangle \leq 1$ [29], in the case of protein dynamics, this condition can be relaxed up to $Q^2\langle u^2\rangle < 4.0$ [30]. It is worth noting that the $\langle u^2\rangle$ values obtained from the experiments are those observed within the time window, which is defined by the energy resolution of the spectrometer employed (Eq. 2.10). Therefore, the $\langle u^2\rangle$ values extracted from Eq. 2.23 should be regarded as "effective" values depending on the energy resolution but not as absolute values.

Since the Gaussian approximation is valid only in lower-Q regions, not all the data points in the measured EINS curves are utilized, and thus analysis taking into account the non-Gaussian behavior at higher-Q regions is required. So far, several sophisticated models have been developed and employed in different systems. Among those, the latest model [31, 32] is introduced here. In this model, $S(Q,\pm\Delta E)$ is described as

$$S(Q, \pm\Delta E) = S(0,\pm\Delta E)\left(1+\frac{Q^2\langle r^2\rangle}{\beta}\right)^{-\beta}, \tag{2.24}$$

$$p(\lambda,\beta) = \frac{\beta\exp(-\beta\lambda)(\beta\lambda)^{\beta-1}}{\Gamma(\beta)}\ (0<\beta<\infty), \tag{2.25}$$

where $\langle r^2 \rangle$ and β denote the mean-square atomic position fluctuation (MSPF) of the motions and the measure of the motional heterogeneity, respectively. $p(\lambda, \beta)$ describes the distribution of the atomic position fluctuations based on a gamma distribution. This distribution has been shown to reproduce the distribution of amplitudes of atomic motions extracted from the trajectory of the molecular dynamics simulation or the normal-mode analysis of proteins [33]. $\lambda = r^2 / \langle r^2 \rangle$ in Eq. 2.25 is the squared atomic position fluctuation relative to the MSPF. In the limit of $\beta \rightarrow \infty$, the Gaussian approximation (Eq. 2.23) is retrieved. Eq. 2.24 can be applied to all the data points in EINS curves, and thus more detailed information than provided by the Gaussian approximation can be extracted. Using this model, dynamical information on amplitudes of motions such as the MSPF values and the distribution of the atomic position fluctuations is extracted from the EINS curves experimentally obtained.

In addition to the analyses described above, other physical quantities can be estimated from the temperature or pressure dependence of $\langle u^2 \rangle$. For example, the effective force constant [8], which is a measure of the molecular rigidity or resilience, and thermodynamic quantities such as the free energy difference ΔG between the two states where atoms reside [34] can be obtained [35, 36]. Thus, while EINS curves derive from only the elastically scattered neutrons, they contain a wealth of information on the dynamics of bio-macromolecules.

2.2.3.2 Quasi-elastic Neutron Scattering (QENS)

In QENS experiments, the elastic and the quasi-elastic components in Figure 2.6 are measured and analyzed to extract dynamical parameters on a given system. The QENS spectra $S(Q, \omega)$ are measured as a two-dimensional map as a function of Q and ω. In many cases, the two-dimensional map is divided into strips in the Q direction and several resultant spectra, each of which is as shown in Figure 2.6, are analyzed by a fitting approach. On the contrary to EINS, QENS permits to distinguish the kind of motions present in the sample, as for instance those shown in Figure 2.1. For that, the following "phenomenological" equation is employed:

$$\begin{aligned} S(Q, \omega) &= C(Q) S_{local}(Q, \omega) \otimes S_{global}(Q, \omega) \otimes R(Q, \omega) + B(Q) \\ &= C(Q)\left[A_0(Q)\delta(\omega) + \sum_{i=1}^{n} A_i(Q) L_i(Q, \omega)\right] \otimes L_G(Q, \omega) \otimes R(Q, \omega) + B(Q), \end{aligned} \tag{2.26}$$

where $S_{local}(Q, \omega)$ and $S_{global}(Q, \omega)$ arise from the local atomic motions in the bio-macromolecules and from the center-of-mass diffusion (global diffusion) of the entire molecules, respectively. $C(Q)$ is the scaling factor between the measured and the calculated spectra, which includes the Q-dependent Debye-Waller factor describing the vibrations of atoms. $A_0(Q)\delta(\omega)$ is the elastic component. $A_0(Q)$ is called the elastic incoherent structure factor (EISF), which contains the information on the geometry of motions (see below). $R(Q, \omega)$ is the resolution function, e.g. mimicked by the QENS spectra of vanadium, and $B(Q)$ is the background, which arises from fast atomic vibrations and/or experimental configurations. $\otimes$ is the convolution operator. Eq. 2.26 describes the QENS spectra as the sum of an elastic component and n types of quasi-elastic components. Each quasi-elastic component is described by a Lorentzian function $L_i(Q, \omega)$:

$$L_i(Q, \omega) = \frac{1}{\pi} \frac{\Gamma_i(Q)}{\omega^2 + \Gamma_i(Q)^2}, \tag{2.27}$$

where $\Gamma_i(Q)$ is the half width at the half maximum (HWHM) of the i-th Lorentzian function. The amplitudes $A_i(Q)$ are normalized to satisfy the condition $\sum_{i=0}^{n} A_i(Q) = 1$. $L_G(Q, \omega)$ in $S_{global}(Q, \omega)$ is also a Lorentzian function with a width of $\Gamma_G(Q)$. If the global diffusion is suppressed, as is the case with hydrated powder samples, the term $S_{global}(Q, \omega)$ should be neglected. Note that the number n of Lorentzian functions describing the local motions depends on the system investigated and is often equal to 1–3. It is also worth noting that the motions in real biomolecular systems cannot be fully described by a few relaxational components. However, increasing n leads to overfitting of the QENS spectra. This is the reason why the phenomenological approach is employed, where all the visible motions with various relaxational times contained in the system are described by a combination of a few representative motions, i.e. Lorentzian functions.

Next, we describe representative dynamical parameters obtained from the phenomenological fitting. In general, the Q-dependence of $\Gamma_i(Q)$, $\Gamma_G(Q)$, and the EISF is analyzed assuming a physical model describing specific atomic motions to extract the relevant dynamical parameters. Since there are various types of models for the Q-dependent behavior of the width of the Lorentzian functions and the EISF [11], in the following, only representative models are introduced. An often used model to analyze the $\Gamma_i(Q)$ behavior and to describe the local atomic motions is the jump-diffusion model [37]. As shown in Figure 2.7a, this model assumes that an atom remains in a given site for a period of τ_{jump} (the residence time), and then it moves rapidly to another site in a negligible time. The diffusion coefficient during the jump motion is called the jump diffusion coefficient (D_{jump}). The jump distance l is calculated by $l = \sqrt{6D_{jump}\tau_{jump}}$ for the three-dimensional case. For this model, $\Gamma_i(Q)$ is described by

$$\Gamma_i(Q) = \frac{D_{jump}Q^2}{1 + D_{jump}Q^2\tau_{jump}}. \tag{2.28}$$

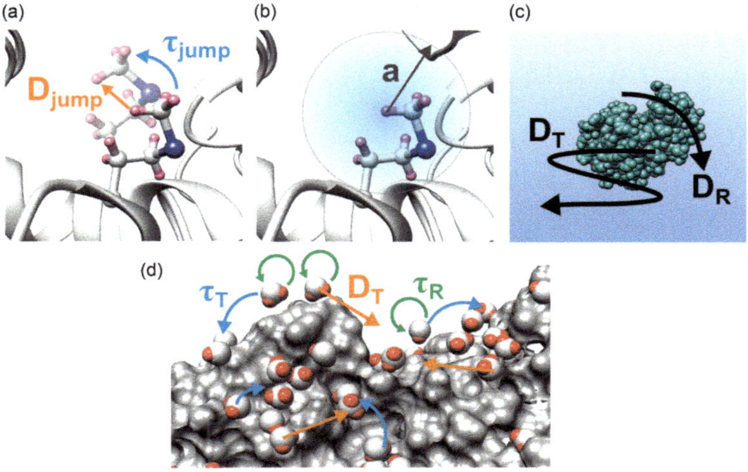

Figure 2.7 Atomic motions observed by incoherent neutron scattering for proteins (a–c) and bulk or hydration water molecules (d). (a, b): Local atomic motions within the protein molecules. Hydrogen atoms and a sulfur atom are shown in hot pink and blue, respectively. (c): Global motions, i.e. the center-of-mass diffusion, of the protein molecules. (d): Water molecules are represented by spheres, where oxygen and hydrogen atoms are depicted in white and red, respectively. Protein surface is depicted in the surface representation. For details of each dynamical parameter, readers are referred to the main text.

If $\Gamma_i(Q)$ is Q-independent, this indicates that the corresponding motion is the jump diffusion among several sites, such as the methyl group rotation [38]. In this case, the corresponding residence time is calculated to be $\tau_{jump} = 1/\Gamma_i(Q)$.

Now, we focus on the EISF. From Eq. 2.14,

$$\begin{aligned} G_s(\vec{r},t) &= \frac{1}{N}\sum_{i=1}^{N}\langle\delta\{\vec{r}-\vec{R_i}(t)+\vec{R_i}(0)\}\rangle \\ &= \frac{1}{N}\sum\nolimits_{i=1}^{N}\int\langle\delta\{\vec{r}-\vec{r}'+\vec{R_i}(0)\}\delta\{\vec{r}'-\vec{R_i}(t)\}\rangle d\vec{r}'. \end{aligned} \tag{2.29}$$

By taking the limit $t \to \infty$, there will obviously be no correlation between $\vec{R_i}(0)$ and $\vec{R_i}(\infty)$. Therefore, Eq. 2.29 reads

$$G_s(\vec{r},\infty) = \frac{1}{N}\sum\nolimits_{i=1}^{N}\int\langle\delta\{\vec{r}-\vec{r}'+\vec{R_i}(0)\}\rangle\langle\delta\{\vec{r}'-\vec{R_i}(\infty)\}\rangle d\vec{r}'. \tag{2.30}$$

We can split $G_s(\vec{r},t)$ into the time-dependent part $G_s^T(\vec{r},t)$ and the time-independent part $G_s(\vec{r},\infty)$ as

$$G_s(\vec{r},t) = G_s(\vec{r},\infty) + G_s^T(\vec{r},t). \tag{2.31}$$

By substituting Eq. 2.31 into Eq. 2.17,

$$\begin{aligned} S_{inc}(\vec{Q},\omega) &= \frac{1}{2\pi\hbar}\iint G_s(\vec{r},\infty)\exp\{i(\vec{Q}\vec{r}-\omega t)\}d\vec{r}dt \\ &\quad + \frac{1}{2\pi\hbar}\iint G_s^T(\vec{r},t)\exp\{i(\vec{Q}\vec{r}-\omega t)\}d\vec{r}dt \\ &= \frac{1}{2\pi\hbar}\int G_s(\vec{r},\infty)\exp(i\vec{Q}\vec{r})d\vec{r}\int\exp(-i\omega t)dt + S_{inc}^{quasi}(\vec{Q},\omega) \\ &= \frac{1}{2\pi\hbar}\int G_s(\vec{r},\infty)\exp(i\vec{Q}\vec{r})d\vec{r}\delta(\omega) + S_{inc}^{quasi}(\vec{Q},\omega), \end{aligned} \tag{2.32}$$

where the first term is the elastic component, and the second term is the quasi-elastic component. The EISF $A_0(Q)$ thus reads

$$\begin{aligned} A_0(Q) &= \int G_s(\vec{r},\infty)\exp(i\vec{Q}\vec{r})d\vec{r} \\ &= \int\left\{\frac{1}{N}\sum_{i=1}^{N}\int\left\langle\delta\{\vec{r}-\vec{r}'+\overrightarrow{R_i}(0)\}\right\rangle\left\langle\delta\{\vec{r}'-\overrightarrow{R_i}(\infty)\}\right\rangle d\vec{r}'\right\}\exp(i\vec{Q}\vec{r})d\vec{r} \\ &= \frac{1}{N}\sum_{i=1}^{N}\int\left[\int\left\langle\delta\{\vec{r}-\vec{r}'+\overrightarrow{R_i}(0)\}\right\rangle\exp(i\vec{Q}(\vec{r}-\vec{r}'))d\vec{r}\right] \\ &\quad \left\{\left\langle\delta\{\vec{r}'-\overrightarrow{R_i}(\infty)\}\right\rangle\exp(i\vec{Q}\vec{r}')d\vec{r}'\right\} \\ &= \frac{1}{N}\sum_{i=1}^{N}\int\left\langle\exp(-i\vec{Q}\overrightarrow{R_i}(0))\right\rangle\left\{\left\langle\delta\{\vec{r}'-\overrightarrow{R_i}(\infty)\}\right\rangle\exp(i\vec{Q}\vec{r}')d\vec{r}'\right\} \\ &= \frac{1}{N}\sum_{i=1}^{N}\left\langle\exp(-i\vec{Q}\overrightarrow{R_i}(0))\right\rangle\left\langle\exp(i\vec{Q}\overrightarrow{R_i}(\infty))\right\rangle \\ &= \frac{1}{N}\sum_{i=1}^{N}\left|\left\langle\exp(-i\vec{Q}\overrightarrow{R_i})\right\rangle\right|^2. \end{aligned} \tag{2.33}$$

In the last step of the above calculation, the following relation was used:

$$\left\langle\exp(-i\vec{Q}\overrightarrow{R_i}(0))\right\rangle = \left\langle\exp(i\vec{Q}\overrightarrow{R_i}(\infty))\right\rangle = \left\langle\exp(-i\vec{Q}\overrightarrow{R_i})\right\rangle, \tag{2.34}$$

where $\overrightarrow{R_i}$ is the position of the i-th atom at an arbitrary time. This relation holds because the value of $\exp(-i\vec{Q}\overrightarrow{R_i})$ averaged over all the possible positions of atoms is the same regardless of the time. The EISF is thus the space-Fourier transform of the final distribution of the identical atoms averaged over all the possible positions and atomic species, and thus it contains the information on the geometry of atomic motions in the sample [11, 12].

The Q-dependence of the EISF is often analyzed using a model called the diffusion-inside-a-sphere model [39]. In this framework, the EISF is described as

$$A_0(Q) = (1 - p_0)\left(\frac{3j_1(Qa)}{Qa}\right)^2 + p_0, \tag{2.35}$$

where p_0 is the fraction of "immobile" atoms, whose motion is too slow to be resolved within the energy resolution employed. The first term represents the motions of "mobile" or resolved atoms. $j_1(Qa)$ is the first-order spherical Bessel function of the first kind, and a is the radius of a sphere within which an atom can move as shown in Figure 2.7b.

In solution samples, proteins undergo global diffusive motions characterized by the translational diffusion coefficient (D_T) and the rotational diffusion coefficient (D_R) (Figure 2.7c). The width of the Lorentzian function arising from the continuous diffusion, such as the global diffusion of proteins, is described by

$$\Gamma_G(Q) = D_{app}Q^2, \tag{2.36}$$

where D_{app} is the "apparent" diffusion coefficient, which contains contributions from both D_T and D_R. If the atomic structure of the investigated molecule is available, the D_T and D_R values can be estimated from the software HYDROPRO [40] and one can compare the measured and simulated D_{app} values [26]. For proteins containing flexible regions, deviations between the measured and simulated D_{app} values are observed [41–43], which are attributed to segmental motions of the molecules.

Due to the large incoherent scattering cross-section of hydrogen atoms, the motion of light water (H_2O) can be studied. Since not only bulk water but also hydration water (i.e. the water molecules bound at the surface of biomolecules) plays a crucial role in protein function [44], characterization of water mobility is of critical importance. The QENS spectra of water molecules are described as [11]

$$\begin{aligned} S_{water}(Q,\omega) &= \{C_1(Q)L_T(Q,\omega) + C_2(Q)L_{T+R}(Q,\omega)\} \\ &\quad \otimes R(Q,\omega) + B(Q) \\ &= \left\{C_1(Q)\frac{\Gamma_T(Q)}{\omega^2 + \Gamma_T(Q)^2} + C_2(Q)\frac{\Gamma_T(Q) + \Gamma_R(Q)}{\omega^2 + (\Gamma_T(Q) + \Gamma_R(Q))^2}\right\} \\ &\quad \otimes R(Q,\omega) + B(Q), \end{aligned} \tag{2.37}$$

where $L_T(Q,\omega)$ and $L_{T+R}(Q,\omega)$ denote the Lorentzian functions describing the translational motions and the sum of the translational and rotational motions of the water molecules, respectively. $C_1(Q)$ and $C_2(Q)$ represent the relative contributions of the two Lorentzian functions. The Q-dependence of $\Gamma_T(Q)$, the width of $L_T(Q, \omega)$, is described by the jump diffusion model [37] (note that the jump diffusion model was originally developed for water motions), from which the residence time τ_T and the translational diffusion coefficient D_T are obtained (Figure 2.7d). On the other hand, $\Gamma_R(Q)$ contained in the width of $L_{T+R}(Q, \omega)$ is Q-independent and described by $\Gamma_R(Q) = {\tau_R}^{-1}$, where τ_R is the rotational correlation time of the water molecule. The translational and rotational motions of water molecules can thus be characterized using QENS.

As described in this section, various dynamical parameters related to the amplitudes, the frequency, and the geometry of atomic motions in bio-macromolecules and water molecules can be extracted through the analysis of the iNS spectra.

2.3 Examples of Biological Applications of iNS

In this section, several recent studies on bio-macromolecules and water molecules using iNS are briefly introduced to see how the basic theories and analyses built on the phenomenological equation described above provide dynamical information relevant to the biological functions of a given system.

2.3.1 Dynamical Modulation of Proteins Caused by a Disease-causing Point Mutation

Human cardiac troponin (Tn), which consists of three subunits, TnC, TnI, and TnT, regulates the cardiac muscle force generation in our heart. Ca^{2+} binding to TnC triggers a series of force generation processes. It is known that various point mutations in Tn disturb the regulatory function, causing a disease called familial cardiomyopathy [45]. However, it remained unclear how a point mutation affects the molecular properties of Tn that lead to pathogenesis. In order to gain insights regarding this aspect, the molecular dynamics of Tn was studied using QENS [42]. In this study, the core domain of Tn (Tn-CD), which is a functionally important part of Tn, was used as sample. The QENS spectra of the wild type Tn-CD and of the Tn-CD containing the K247R mutation of TnT taken at 12 μeV energy resolution using the backscattering spectrometer BL02 DNA [46] in J-PARC were analyzed. Note that this mutation is known to increase the maximum force developed by cardiac muscles.

Regarding the global motions, D_{app} (see Eq. 2.36) was found not to be fully explained by the sum of the contributions of D_T and D_R for both samples, indicating that segmental motions contribute to the measured D_{app} values (~30% of residues in Tn-CD are in the disordered state). The contribution of the segmental motions was similar within errors between the wild type and the mutant. On the other hand, from the EISF analysis (Eq. 2.35), it was found that in the Ca^{2+}-binding state, the radius of a sphere where atoms can move is larger for the mutant than for the wild type, suggesting that the atoms of the mutant can explore a larger conformational space than those of the wild type. Thus, the mutation does not affect large-scale motions such as segmental motions but only modulates the local atomic motions, indicating that the regulatory function is controlled by a fine-tuned mechanism. It was also suggested that the larger flexibility of the mutant affects the structural changes of downstream proteins that Tn directly interacts with. Indeed, it was later confirmed that the E244D mutation of TnT, which shows the same functional aberration as K247R, alters the structure of the downstream proteins in the direction to promote force generation, i.e. the functional aberration observed for this mutation [47].

2.3.2 Dynamical Differences between Amyloid Polymorphic Fibrils Showing Different Levels of Cytotoxicity

Lysozyme amyloidosis is a rare but severe hereditary disease, which is characterized by the deposition of amyloid fibrils in several internal organs. It is known that amyloid fibrils show polymorphism, where fibrils take on different morphologies depending on fibrillation conditions. The polymorphs of hen egg white lysozyme (HEWL), which has often been used as a model of human lysozyme, have been shown to express different levels of cytotoxicity [48]. However, the molecular mechanism of cytotoxicity and thus lysozyme amyloidosis remains unclear. Recent studies have shown that the binding of amyloid fibrils to cell membranes triggers a series of processes leading to cytotoxicity and that the surface structure of fibrils contributes to cytotoxicity [49]. Since protein flexibility in general plays a crucial role in binding processes, it is important to reveal possible differences in dynamics between polymorphs, which show different levels of cytotoxicity, to ultimately understand the molecular mechanism of cytotoxicity.

The dynamics of two kinds of HEWL polymorphs, which are known to show high and low levels of cytotoxicity, was investigated at 8 μeV energy resolution using EINS and QENS [50] on the backscattering spectrometer IN13 [51] at ILL. The former and the latter polymorphs are called LP_{high} and LP_{low}, respectively. An analysis of the EINS data revealed that while the MSD value is similar between the two samples, the β value in Eq. 2.24 is significantly different, suggesting that the dynamical difference lies in the distribution of amplitudes. LP_{high} contains a larger fraction of atoms moving with larger amplitudes than LP_{low}. Moreover, QENS analysis revealed that D_{jump} of atoms in LP_{high} is larger than that in LP_{low}, indicating that the atoms in LP_{high} undergo more rapid motions than those in LP_{low}. This study thus revealed that the dynamics of LP_{high} is enhanced compared with LP_{low} at the atomic level. Since enhanced conformational fluctuation generally promotes biochemical reactions [52, 53], LP_{high} is likely to accelerate the binding of fibrils to cell membranes, eventually causing a higher level of cytotoxicity.

2.3.3 New Theoretical Framework to Describe the Dynamical Behavior of Lipid Molecules

Lipid molecules form cell membranes and play important roles in various cell functions such as a scaffold for membrane proteins. Since the flexible nature and fluidity of lipid membranes are considered critical for lipid functions, the dynamics of lipid molecules have been investigated using iNS. In addition to experimental studies, considerable efforts have been made to develop theoretical models that can interpret the QENS data and extract dynamical information in detail. The first model developed by Pfeiffer et al. in 1989 [54] succeeded to decipher several motions that lipid molecules undergo in bilayer membranes. However, due to the limitation on the neutron flux and the QENS data quality, it was not possible to employ many free parameters in the model in order to avoid overfitting. Since then, a rise in neutron flux and upgrades of spectrometers have permitted to obtain neutron scattering data with higher quality. Therefore, a more detailed and generic framework for lipid dynamics has been required. A novel model in this direction has been developed by Wanderlingh et al. [55] in combination with the analysis of the high-quality QENS data, where the spectra were divided into three Lorentzian functions with each having a significantly different energy width.

Recently, Bicout et al. have further developed a new comprehensive dynamical model named "Matryoshka model" to describe various motions contained in phospholipid molecules based on the corresponding QENS spectra [56–58]. In this model, seven kinds of motions occurring at different timescales ranging from global diffusion of lipid molecules to local atomic motions such as the trans-gauche isomerization of methylene groups are considered. By incorporating all the contributions, the analytical expression of the scattering intensity $S_{Matryoshka}(Q, \omega)$ of lipid molecules was derived. As an illustrative application, the Matryoshka model was applied to analyze the QENS spectra of DMPC (1,2-dimyristoyl-*sn*-glycero-3-phosphocholine) in multilamellar bilayers or vesicles. The QENS spectra were decomposed into three Lorentzian functions (see Eq. 2.26) with significantly different widths, which are classified as "slow," "intermediate," and "fast" motions, in addition to the elastic component [55]. The analytical expression of the scattering intensity for the three classes of motions was derived from $S_{Matryoshka}(Q, \omega)$ to extract dynamical parameters. It was found that the Q-dependence of the amplitudes $A_i(Q)$ and the widths Γ_i (Q) of the Lorentzian functions is satisfactorily reproduced by this model [57, 58], suggesting that the Matryoshka model can reproduce not only the geometry of motions but also the rates of atomic motions contained in the lipid molecules. Except for a few dynamical parameters where their $A_i(Q)$ profiles show similar behavior and hence are difficult to be separated, the majority of the parameters assumed in this model can be determined with error values that are typically 5–10% of the obtained parameter values, allowing to distinguish the dynamical processes between the head and the tail groups, between the samples, and between the temperatures. Furthermore, in this model, overfitting is avoided. Overfitting generally produces fitted curves that go through all the data points unnaturally within their errors. Whereas the Matryoshka model reproduces the experimental Ai(Q) profiles quite well as a whole, they deviate from several data points [57], implying that the model contains a smaller number of variables than that required for overfitting. In addition, $A_i(Q)$ profiles of the model share the same parameters. In that way, the Matryoshka model can be applied in a global way to all the amplitudes, which constrains the fit, reduces overfitting, and increases the precision of the calculations. Despite the fact that the Matryoshka model does not include the temperature dependence of the dynamical parameters, enhanced dynamics is revealed at higher temperature as expected. The extracted dynamical parameters of the head group and the tail group in DMPC molecules are in agreement with the findings in the literature [59]. Thus, the Matryoshka model is a state-of-the-art theoretical framework to describe the lipid dynamics and will demonstrate its power to characterize more complex systems such as natural cell membranes.

2.3.4 Separation of Dynamics of Protein-detergent Complexes

In the above examples, the samples contain only one type of bio-macromolecules. On the other hand, combined with elaborated analytical technique, the dynamics of more complex systems such as protein-detergent complexes can be studied. Apolipoprotein B-100 (apo B-100) is a protein composed of low-density lipoproteins (LDL) and very-low-density lipoproteins (VLDL), which play a major role in the transport of lipids such as triglycerides and cholesterol. Apo B-100 is involved in the binding to cellular lipoprotein receptors. VLDL is converted to LDL during metabolism, resulting in a rearrangement of apo B-100 utilizing its flexible structural features. An EINS study on LDL and VLDL has shown that the MSD values are significantly different between them. From this observation, the authors suggested that apo B-100 shows different dynamics dependent on the lipoprotein that it is bound to [60]. It is thus important to study the molecular dynamics of apo B-100, especially in an isolated state, to understand the mechanism of lipid metabolism in detail. On the other hand, since apo B-100 is an amphiphilic protein, it requires detergents to obtain solubilized form of isolated apo B-100, which makes the dynamical analysis by iNS more challenging.

Recently, the dynamics of apo B-100 in complex with detergents called Nonidet P-40 (NP40) was successfully analyzed by separating both dynamical contributions in the QENS spectra [61]. The QENS spectra of hydrated powder samples of apo B-100/NP40 complexes were recorded at multiple energy resolutions of 10 μeV, 4 μeV, and 0.75 μeV on the instruments IN5 [62], IN16B, and IN16B-BATS [63] at ILL in order to separate and analyze the dynamical features of each component. The QENS spectra were deconvoluted into two or three Lorentzian functions in addition to the elastic component. The assignment of various motions contained in the apo B-100/NP40 complexes to each Lorentzian function was done based on a newly developed dynamical model, which was inspired by the Matryoshka model described in 2.3.3. From the analysis of the width of the Lorentzian functions, it was found that the residence time τ of domain motions of apo B-100 is about 10 times larger than that of NP40 at 0.75 μeV (~1 ns). Moreover, the analysis of the Q-dependence of $A_i(Q)$ (see Eq. 2.26) showed that the radius of a sphere where atoms can move is smaller for apo B-100 than for NP40 within the same time window. These results suggest that the domain motions of apo B-100 are highly confined through interactions with NP40 in apo B-100/NP40 complexes. This study thus opened up a new possibility to decipher the dynamics of a target molecule in complex systems such as membrane proteins embedded in lipid bilayers.

2.3.5 Hydration Water Mobility around Proteins

As described in a previous section, iNS is suitable to characterize the water mobility as well as dynamics of bio-macromolecules. In general, the mobility of hydration water around proteins is decreased due to the interaction between water molecules and the residues on the protein surface [44]. However, for proteins called fibrous actin (F-actin), hydration water whose rotational mobility is close to that of bulk water has been identified using dielectric spectroscopy [64, 65]. Such hydration water is called hyper-mobile water. F-actin is a polymerized form of monomeric actin and is involved in many cellular functions. One of its functions is the force development in muscles as a result of the interaction between F-actin and its partner protein called myosin. The above study using dielectric spectroscopy identified the hyper-mobile water only for F-actin while the "normal" hydration water, whose mobility is reduced compared with bulk water, was detected for myosin.

In order to characterize further the dynamical properties of the hydration water of F-actin (HW_{Fa}) and that of myosin (HW_{myo}), QENS spectra were measured on solution samples of F-actin and myosin [66] at 26.6 μeV and 90.5 μeV using the time-of-flight spectrometer BL14 AMATERAS in J-PARC [67]. First, the QENS spectra arising only from proteins were obtained by subtracting the spectra of D_2O solvent from those of the samples containing both proteins and D_2O solvent [68]. Next, the QENS spectra corresponding to the sum of bulk and hydration water were obtained by subtracting the spectra of proteins from those of the samples containing both proteins and H_2O solvent. The remaining spectra were then separated into the spectra arising from bulk water and from hydration water, assuming two hydration layers around proteins [69]. The resultant spectra of hydration water were analyzed using Eq. 2.37. It was found that the rotational correlation time (τ_R) of HW_{Fa} is similar to that of bulk water, which is in agreement with the results of the dielectric spectroscopy studies. Regarding the translational motions, whereas the residence times (τ_T) are similar within errors between

HW_{Fa} and HW_{myo}, the translational diffusion coefficient (D_T) of HW_{Fa} is much larger than that of HW_{myo} and is even close to that of bulk water. Thus, the QENS study has shown that both the translational and rotational mobility of the hydration water around F-actin is enhanced compared with the typical protein hydration water. This finding will serve as a basis to understand the interactions between F-actin and many actin-binding proteins.

2.4 Summary

This chapter described the basic theory and fundamental knowledge obtained by incoherent neutron scattering (iNS), aiming not to use complex mathematics and knowledge in quantum physics. So far, sophisticated iNS studies have often been carried out on model proteins or lipids such as myoglobin and DMPC to profoundly understand the atomic behavior underlying the biological functions. At the same time, more physiologically and biologically important bio-macromolecules have been identified and investigated in the fields of molecular biology, biochemistry, and physiology. Application of iNS to these samples and the subsequent integration of the findings obtained from these different fields will thus help to establish a link between atomic-scale information and macroscopic phenomena not only in a physiological but also in a pathological context. Furthermore, in addition to the existing research reactors and spallation sources, another large facility named European Spallation Source (ESS) in Sweden is under construction (https://europeanspallationsource.se). Thus, there will be more opportunities to carry out iNS experiments in the near future. Readers interested in carrying out iNS experiments on bio-macromolecules could consult the websites of neutron facilities or directly contact the authors of this chapter.

References

1 Henzler-Wildman, K.A., Thai, V., Lei, M. et al. (2007). "Intrinsic motions along an enzymatic reaction trajectory." *Nature* 450: 838–844. doi: 10.1038/nature06410.

2 Wakabayashi, K., Tokunaga, M., Kohno, I. et al. (1992). "Small-angle synchrotron x-ray scattering reveals distinct shape changes of the myosin head during hydrolysis of ATP." *Science* (80-) 258: 443–447.

3 Makowski, L. (2010). "Characterization of proteins with wide-angle X-ray solution scattering (WAXS)." *J Struct Funct Genomics* 11: 9–19. doi: 10.1007/s10969-009-9075-x.

4 Peters, J., Martinez, N., Trovaslet, M. et al. (2016). "Dynamics of human acetylcholinesterase bound to non-covalent and covalent inhibitors shedding light on changes to the water network structure." *Phys Chem Chem Phys* 18: 12992–13001. doi: 10.1039/c6cp00280c.

5 Smith, J.C. (1991). "Protein dynamics: Comparison of simulations with inelastic neutron scattering experiments." *Q Rev Biophys* 24: 227–291. doi: 10.1017/S0033583500003723.

6 Henzler-Wildman, K.A., Lei, M., Thai, V. et al. (2007). "A hierarchy of timescales in protein dynamics is linked to enzyme catalysis." *Nature* 450: 913–916. doi: 10.1038/nature06407.

7 Henzler-Wildman, K. and Kern, D. (2007). "Dynamic personalities of proteins." *Nature* 450: 964–972.

8 Zaccai, G. (2000). "How soft is a protein? A protein dynamics force constant measured by neutron scattering." *Science* (80-) 288: 1604–1607. doi: 10.1126/science.288.5471.1604.

9 Gabel, F., Bicout, D., Lehnert, U. et al. (2002). "Protein dynamics studied by neutron scattering." *Q Rev Biophys* 35: 327–367. doi: 10.1017/s0033583502003840.

10 Bacon, G. (1975). *Neutron Diffraction*. London: Oxford University Press.

11 Bée, M. (1988). *Quasielastic Neutron Scattering*. Bristol and Philadelphia: Adam Hilger.

12 Squires, G.L. (1978). *Introduction to the Theory of Thermal Neutron Scattering*. Cambridge: Cambridge University Press.

13 Fitter, J., Gutberlet, T., and Katsaras, J. (2006). *Neutron Scattering in Biology*. Springer-Verlag Berlin Heidelberg.

14 Qian, S., Sharma, V.K., and Clifton, L.A. (2020). "Understanding the structure and dynamics of complex biomembrane interactions by neutron scattering techniques." *Langmuir* 36: 15189–15211. doi: 10.1021/acs.langmuir.0c02516.

15 Gell, Y. and Lichtenberg, D.B. (1969). "Quark model and the magnetic moments of proton and neutron." *Nuovo Cim A* 61: 27–40. doi: 10.1007/BF02760010.

16 Jacrot, B. and Zaccai, G. (1981). "Determination of molecular weight by neutron scattering." *Biopolymers* 20: 2413–2426. doi: 10.1002/bip.1981.360201110.

17 Sears, V.F. (1992). "Neutron scattering lengths and cross sections." *Neutron News* 3: 26–37. doi: 10.1080/10448639208218770.

18 Svergun, D.I., Richard, S., Koch, M.H. et al. (1998). "Protein hydration in solution: Experimental observation by x-ray and neutron scattering." *Proc Natl Acad Sci U S A* 95: 2267–2272.

19 Matsuo, T. (2020). "Usefulness of medium-angle X-ray scattering for structural characterization of flexible proteins studied by computer simulations." *Biochem Biophys Res Commun* 525: 830–835. doi: 10.1016/j.bbrc.2020.02.150.

20 Gaspar, A.M., Busch, S., Appavou, M.S. et al. (2010). "Using polarization analysis to separate the coherent and incoherent scattering from protein samples." *Biochim Biophys Acta* 1804: 76–82. doi: 10.1016/j.bbapap.2009.06.024.

21 Matsuo, T., Arluison, V., Wien, F., and Peters, J. (2022). "Structural information on bacterial amyloid and amyloid-DNA complex obtained by small-angle neutron/X-ray scattering." In: *Methods in Molecular Biology "Bacterial Amyloids"*. Edited by, Véronique Arluison, Frank Wien, Andrés Marcoleta Springer, Humana Press 6. vol. 2538, pp. 95–107.

22 Richard, D., Ferrand, M., and Kearley, G.J. (1996). "Analysis and visualisation of neutron-scattering data." *J Neutron Res* 4: 33–39. doi: 10.1080/10238169608200065.
23 Azuah, R.T., Kneller, L.R., Qiu, Y. et al. (2009). "DAVE: A comprehensive software suite for the reduction, visualization, and analysis of low energy neutron spectroscopic data." *J Res Natl Inst Stand Technol* 114: 341–358. doi: 10.6028/jres.114.025.
24 Inamura, Y., Nakatani, T., Suzuki, J., and Otomo, T. (2013). "Development status of software "Utsusemi" for chopper spectrometers at MLF, J-PARC." *J Phys Soc Japan* 82: SA031. doi: 10.7566/jpsjs.82sa.sa031.
25 Arnold, O., Bilheux, J.C., Borreguero, J.M. et al. (2014). "Mantid—Data analysis and visualization package for neutron scattering and μ SR experiments." *Nucl Instruments Methods Phys Res Sect A Accel Spectrometers, Detect Assoc Equip* 764: 156–166. doi: 10.1016/j.nima.2014.07.029.
26 Perez, J., Zanotti, J.M., and Durand, D. (1999). "Evolution of the internal dynamics of two globular proteins from dry powder to solution." *Biophys J* 77: 454–469. doi: 10.1016/S0006-3495(99)76903-1.
27 Gaspar, A.M., Appavou, M.S., Busch, S. et al. (2008). "Dynamics of well-folded and natively disordered proteins in solution: A time-of-flight neutron scattering study." *Eur Biophys J* 37: 573–582. doi: 10.1007/s00249-008-0266-3.
28 Rahman, A., Singwi, K., and Sjölander, A. (1962). "Theory of Slow Neutron Scattering by Liquids." *I Phys Rev* 126: 986–996. doi: 10.1103/PhysRev.126.986.
29 Gabel, F. (2005). "Protein dynamics in solution and powder measured by incoherent elastic neutron scattering: The influence of Q-range and energy resolution." *Eur Biophys J* 34: 1–12. doi: 10.1007/s00249-004-0433-0.
30 Stadler, A.M., Unruh, T., Namba, K. et al. (2013). "Correlation between supercoiling and conformational motions of the bacterial flagellar filament." *Biophys J* 105: 2157–2165. doi: 10.1016/j.bpj.2013.09.039.
31 Peters, J. and Kneller, G.R. (2013). "Motional heterogeneity in human acetylcholinesterase revealed by a non-Gaussian model for elastic incoherent neutron scattering." *J Chem Phys* 139: 165102. doi: 10.1063/1.4825199.
32 Zeller, D., Telling, M.T.F., Zamponi, M. et al. (2018). "Analysis of elastic incoherent neutron scattering data beyond the Gaussian approximation." *J Chem Phys* 149: 234908. doi: 10.1063/1.5049938.
33 Kneller, G.R. and Hinsen, K. (2009). "Quantitative model for the heterogeneity of atomic position fluctuations in proteins: A simulation study." *J Chem Phys* 131: 45104. doi: 10.1063/1.3170941.
34 Bicout, D.J. and Zaccai, G. (2001). "Protein flexibility from the dynamical transition: A force constant analysis." *Biophys J* 80: 1115–1123. doi: 10.1016/S0006-3495(01)76089-4.
35 Peters, J., Marion, J., Natali, F. et al. (2017). "The dynamical transition of lipid multilamellar bilayers as a matter of cooperativity." *J Phys Chem B* 121: 6860–6868. doi: 10.1021/acs.jpcb.7b05167.

36 Cisse, A., Marquette, A., Altangerel, M. et al. (2021). "Investigation of the action of peptides on lipid membranes." *J Phys Chem B* 125: 10213–10223. doi: 10.1021/acs.jpcb.1c06388.

37 Singwi, K. and Sjölander, A. (1960). "Diffusive motions in water and cold neutron scattering." *Phys Rev* 119: 863–871. doi: 10.1103/PhysRev.119.863.

38 Roh, J.H., Novikov, V.N., Gregory, R.B. et al. (2005). "Onsets of anharmonicity in protein dynamics." *Phys Rev Lett* 95: 1–4. doi: 10.1103/PhysRevLett.95.038101.

39 Volino, F. and Dianoux, A.J. (1980). "Neutron incoherent scattering law for diffusion in a potential of spherical symmetry: General formalism and application to diffusion inside a sphere." *Mol Phys* 41: 271–279. doi: 10.1080/00268978000102761.

40 García De La Torre, J., Huertas, M.L., and Carrasco, B. (2000). "Calculation of hydrodynamic properties of globular proteins from their atomic-level structure." *Biophys J* 78: 719–730.

41 Stadler, A.M., Stingaciu, L., Radulescu, A. et al. (2014). "Internal nanosecond dynamics in the intrinsically disordered myelin basic protein." *J Am Chem Soc* 136: 6987–6994. doi: 10.1021/ja502343b.

42 Matsuo, T., Tominaga, T., Kono, F. et al. (2017). "Modulation of the picosecond dynamics of troponin by the cardiomyopathy-causing mutation K247R of troponin T observed by quasielastic neutron scattering." *Biochim Biophys Acta Proteins Proteom* 1865: 1781–1789. doi: 10.1016/j.bbapap.2017.09.007.

43 Fujiwara, S., Matsuo, T., Sugimoto, Y., and Shibata, K. (2019). "Segmental motions of proteins under non-native states evaluated using quasielastic neutron scattering." *J Phys Chem Lett* 10: 7505–7509. doi: 10.1021/acs.jpclett.9b03196.

44 Ball, P. (2008). "Water as an active constituent in cell biology." *Chem Rev* 108: 74–108. doi: 10.1021/cr068037a.

45 Harada, K. and Morimoto, S. (2004). "Inherited cardiomyopathies as a troponin disease." *Jpn J Physiol* 54: 307–318.

46 Shibata, K., Takahashi, N., Kawakita, Y. et al. (2015). "The performance of TOF near backscattering spectrometer DNA in MLF, J-PARC." *JPS Conf Proc* 8: 36022. doi: 10.7566/jpscp.8.036022.

47 Matsuo, T., Kono, F., and Fujiwara, S. (2019). "Effects of the cardiomyopathy-causing E244D mutation of troponin T on the structures of cardiac thin filaments studied by small-angle X-ray scattering." *J Struct Biol* 205: 196–205. doi: 10.1016/j.jsb.2018.12.005.

48 Mocanu, M.M., Ganea, C., Siposova, K. et al. (2014). "Polymorphism of hen egg white lysozyme amyloid fibrils influences the cytotoxicity in LLC-PK1 epithelial kidney cells." *Int J Biol Macromol* 65: 176–187. doi: 10.1016/j.ijbiomac.2014.01.030.

49 Yoshiike, Y., Akagi, T., and Takashima, A. (2007). "Surface structure of amyloid-β fibrils contributes to cytotoxicity." *Biochemistry* 46: 9805–9812. doi: 10.1021/bi700455c.

50 Matsuo, T., Francesco De, A., and Peters, J. (2022). "Molecular dynamics of lysozyme amyloid polymorphs studied by incoherent neutron scattering." *Front Mol Biosci* 8: 812096. doi: 10.3389/fmolb.2021.812096.

51 Natali, F., Peters, J., Russo, D. et al. (2008). "IN13 backscattering spectrometer at ILL: Looking for motions in biological macromolecules and organisms." *Neutron News* 19: 14–18. doi: 10.1080/10448630802474083.

52 Shoemaker, B.A., Portman, J.J., and Wolynes, P.G. (2000). "Speeding molecular recognition by using the folding funnel: The fly-casting mechanism." *Proc Natl Acad Sci U S A* 97: 8868–8873. doi: 10.1073/pnas.160259697.

53 Kuroi, K., Okajima, K., Ikeuchi, M. et al. (2014). "Transient conformational fluctuation of TePixD during a reaction." *Proc Natl Acad Sci U S A* 111: 14764–14769. doi: 10.1073/pnas.1413222111.

54 Pfeiffer, W., Henkel, T., Sackmann, E. et al. (1989). "Local dynamics of lipid bilayers studied by incoherent quasi-elastic neutron scattering." *Europhys Lett* 8: 201–206. doi: 10.1209/0295-5075/8/2/016.

55 Wanderlingh, U., D'Angelo, G., Branca, C. et al. (2014). "Multi-component modeling of quasielastic neutron scattering from phospholipid membranes." *J Chem Phys* 140: 174901. doi: 10.1063/1.4872167.

56 Bicout, D.J., Cisse, A., Matsuo, T., and Peters, J. (2022). "The dynamical Matryoshka model: 1. Incoherent neutron scattering functions for lipid dynamics in bilayers." *Biochim Biophys Acta – Biomembr* 1864: 183944. doi: 10.1016/j.bbamem.2022.183944.

57 Cissé, A., Matsuo, T., Plazanet, M. et al. (2022). "The dynamical Matryoshka model: 2. A new modelling of local lipid dynamics at the sub-nanosecond timescale in phospholipid membranes." *BBA – Biomembr* 1864: 183950. doi: 10.1016/j.bbamem.2022.183950.

58 Matsuo, T., Cisse, A., Plazanet, M. et al. (2022). "The dynamical Matryoshka model : 3. Diffusive nature of the atomic motions contained in a new dynamical model for deciphering local lipid dynamics." *BBA – Biomembr* 1864: 183949. doi: 10.1016/j.bbamem.2022.183949.

59 Aoun, B., Pellegrini, E., Trapp, M. et al. (2016). "Direct comparison of elastic incoherent neutron scattering experiments with molecular dynamics simulations of DMPC phase transitions." *Eur Phys J E* 39. doi: 10.1140/epje/i2016-16048-y.

60 Mikl, C., Peters, J., Trapp, M. et al. (2011). "Softness of atherogenic lipoproteins: A comparison of very low density lipoprotein (VLDL) and low density lipoprotein (LDL) using elastic incoherent neutron scattering (EINS)." *J Am Chem Soc* 133: 13213–13215. doi: 10.1021/ja203679g.

61 Cisse, A., Schachner-Nedherer, A.-L., Appel, M. et al. (2021). "Dynamics of apolipoprotein B-100 in interaction with detergent probed by incoherent neutron scattering." *J Phys Chem Lett*: 12402–12410. doi: 10.1021/acs.jpclett.1c03141.

62 Jacques, O., Mutka, H., and Didier, L. (2010). "The new cold neutron time-of-flight spectrometer IN5." *Neutron News* 21: 22–25. doi: 10.1080/10448631003757573.

63 Appel, M., Frick, B., and Magerl, A. (2019). "First results with the neutron backscattering and TOF spectrometer option BATS on IN16B." *Phys B Condens Matter* 562: 6–8. doi: 10.1016/j.physb.2018.11.062.

64 Kabir, S.R., Yokoyama, K., Mihashi, K. et al. (2003). "Hyper-mobile water is induced around actin filaments." *Biophys J* 85: 3154–3161. doi: 10.1016/S0006-3495(03)74733-X.

65 Suzuki, M., Kabir, S.R., Siddique, M.S. et al. (2004). "Myosin-induced volume increase of the hyper-mobile water surrounding actin filaments." *Biochem Biophys Res Commun* 322: 340–346. doi: 10.1016/j.bbrc.2004.07.111.

66 Matsuo, T., Arata, T., Oda, T. et al. (2016). "Difference in the hydration water mobility around F-actin and myosin subfragment-1 studied by quasielastic neutron scattering." *Biochem Biophys Reports* 6: 220–225. doi: 10.1016/j.bbrep.2016.04.013.

67 Nakajima, K., Ohira-Kawamura, S., Kikuchi, T. et al. (2011). "AMATERAS: A cold-neutron disk chopper spectrometer." *J Phys Soc Japan* 80: SB028. doi: 10.1143/jpsjs.80sb.sb028.

68 Matsuo, T., Arata, T., Oda, T. et al. (2015). "Internal dynamics of F-actin and myosin subfragment-1 studied by quasielastic neutron scattering." *Biochem Biophys Res Commun* 459: 493–497. doi: 10.1016/j.bbrc.2015.02.134.

69 Perticaroli, S., Ehlers, G., Stanley, C.B. et al. (2017). "Description of hydration water in protein (green fluorescent protein) solution." *J Am Chem Soc* 139: 1098–1105. doi: 10.1021/jacs.6b08845.

70 Pettersen, E.F., Goddard, T.D., Huang, C.C. et al. (2004). "UCSF Chimera–a visualization system for exploratory research and analysis." *J Comput Chem* 25: 1605–1612. doi: 10.1002/jcc.20084.

3

Elucidation of Protein Function Using Raman Spectroscopy

Saima Malik, Maitrayee U. Trivedi, Gurpreet K. Soni, and Rohit K. Sharma*

Department of Chemistry & Centre for Advanced Studies in Chemistry, Panjab University, Chandigarh, India
** Corresponding author*

3.1 Introduction

In the living world, proteins occupy the topmost position due to their irreplaceable and countless properties. Almost every function and property that exemplifies a living organism is associated with proteins. It is worth mentioning that a living being has almost 10^5 proteins [1]. All living organisms exhibit a variety of complex chemical reactions happening inside the body, which play a central role in life. Proteins are known to catalyze various biochemical reactions [2]. The primary structure of proteins is defined by a sequence of amino acids adjoined by covalent peptide bonds, wherein the secondary structure refers to the local folding pattern of the polypeptide backbone stabilized by hydrogen bonds between N—H and C=O groups. Common secondary structures observed are the α-helix, the β-sheet, loops, turns, etc. [3]. These structures accumulate to form the three-dimensional structure of proteins. Misfolding of the proteins further brings numerous unidentified facts regarding the aggregate formation and toxicity. To analyze the function and the structure of proteins, efficient methods such as nuclear magnetic resonance (NMR) spectroscopy, Fourier-transform infrared (FTIR) spectroscopy [4] and X-ray diffraction have been vigorously explored. However, all these techniques have limitations. Thus, another technique such as Raman spectroscopy has been widely used for the structure determination of the proteins. C.V. Raman made a major contribution to what is currently called Raman spectroscopy in

---

One chapter for a book entitled “Emerging techniques for functional analysis of protein” May 10th 2022

Analytical Techniques for the Elucidation of Protein Function, First Edition. Edited by Isao Suetake, Rohit K. Sharma, and Hironobu Hojo.

1928, and thereafter it has been extensively studied in the scientific community [5]. Its user-friendly applications such as its need for very small amounts with less time for analysis make it promising and attracts interest for industrial applications. In addition, the Raman spectrum can determine the specific signatures of proteins within the complex biosystems such as bacteria and viruses [6].

The advantages of Raman spectroscopy for exploring the structure of proteins are listed below:

1) The method is widely used irrespective of sample morphology, and thus for proteins, whether suspension, precipitates, gels, films, fibers, single crystals, and polycrystalline and amorphous solids, it can be considered. This is a beneficial point for comparing the obtained data of the same protein with the different morphological states.
2) When the sample is dissolved in co-solvent of normal water (H_2O) and heavy water (D_2O), it generates very weak Raman spectra, thus producing relatively little interference. This is a significant advantage over FTIR absorption spectroscopy, where both H_2O and D_2O are highly problematic solvents and interfere in the signal recording. Thus, the low interference from water facilitates the exploitation of the hydrogen-deuterium exchange processes in proteins.
3) A small amount of sample is sufficient for the analysis by surface-enhanced Raman spectroscopy (SERS), which is in marked contrast to other structural methods, including X-ray crystallography and NMR spectroscopy.
4) Laser Raman spectroscopy requires no labels or probes and is nondestructive to protein samples. Accordingly, the native structure can be directly analyzed and the protein may be recovered.
5) The polarized Raman methods can be employed to determine the directionality of the protein main chain within the lattice or to determine configurations of specific side chains.
6) Raman spectroscopy requires less time ($=10^{-15}$ s) for analysis compared to that of fluorescence ($> 10^{-9}$ s) and NMR phenomena ($=10^{-6}$ s).
7) Raman intensities are enhanced dramatically (several orders of magnitude) when the energy of the incident photon is selected in resonance with a molecular electronic transition of a protein chromophore. Thus, the structural information about the chromophore can be obtained in very dilute protein solutions using Raman spectroscopy and thus display valuable advantages for proteins containing chromophores that absorb in the visible (metalloproteins, retinal-proteins, etc.) or near-ultraviolet (nucleotide-binding proteins, nucleoprotein complexes, etc.).
8) There exists a large database of Raman spectra of polypeptides, amino acids, and related model compounds for which reliable band assignments, normal mode analyses, and spectra-structure correlations have been made, which further facilitates the interpretation of the complex Raman spectra obtained from proteins and their assemblies.

3.2 Basic Principle and Working of Raman Spectroscopy

The basic principle and working of Raman spectroscopy involve the scattering of light in an inelastic way, where the quanta of energy are transferred from a beam of monochromatic radiation (laser) to protein and further to beam infringes. From the incident light, only a small fraction of the incident photons is inelastically scattered, and the majority are scattered elastically (Rayleigh scattering) or quasi-elastically (Brillouin scattering). Furthermore, the quantity of energy transferred in each Raman event is only a small fraction of the total energy of the incident photon. Afterward, the quanta are prolonged by the protein in the form of vibrational energies. These discrete vibrational energies transferred in these processes (Raman frequencies), as well as their relative scattering probabilities (Raman intensities) and tensor characteristics (Raman polarization), comprise the vibrational Raman spectrum of the protein. The spectrum is a unique and sensitive function of the intramolecular geometry and intra- and intermolecular force fields. Thus, the spectrum of a protein obtained varies depending on its energy and environment.

The Raman spectrum of a protein is attained experimentally by measuring the energies (frequencies) and relative numbers (intensities) of photons that have been scattered by the target protein. The scattered photons exhibiting frequencies less than the frequency of the impingent laser photons reflect the vibrational quanta transferred to the protein and thus identify the (Stokes) Raman frequencies of the protein.

In general, the spectrum is displayed as a series of bands, which means the Raman intensities are plotted as a function of the scattered frequencies, which are usually expressed in wavenumber (cm^{-1}) units and the intensities in arbitrary units. For additional understandable context, the frequencies of incident and scattered photons are clustered within a relatively narrow region of the electromagnetic spectrum and are summarized in a diagrammatic way (Figure 3.1a).

3.2.1 Theory and Frequencies of Raman Spectroscopy

As we know, for a protein of N atoms, 3N − 6 normal modes of vibration are possible. In principle, each of the 3N − 6 modes extend over the entire molecule, and because the whole protein molecule lacks symmetry, each is capable of leading to a Raman transition. In practice, however, most of the modes are highly localized, meaning that, to a good approximation, each represents a vibration of a small group of atoms largely isolated from the vibrations of other molecular subgroups. In a protein when frequencies are identifiable, they can be recognized as group frequencies. Many group frequencies of the protein main chain and of various side chains have been identified on the basis of extensive Raman and infrared

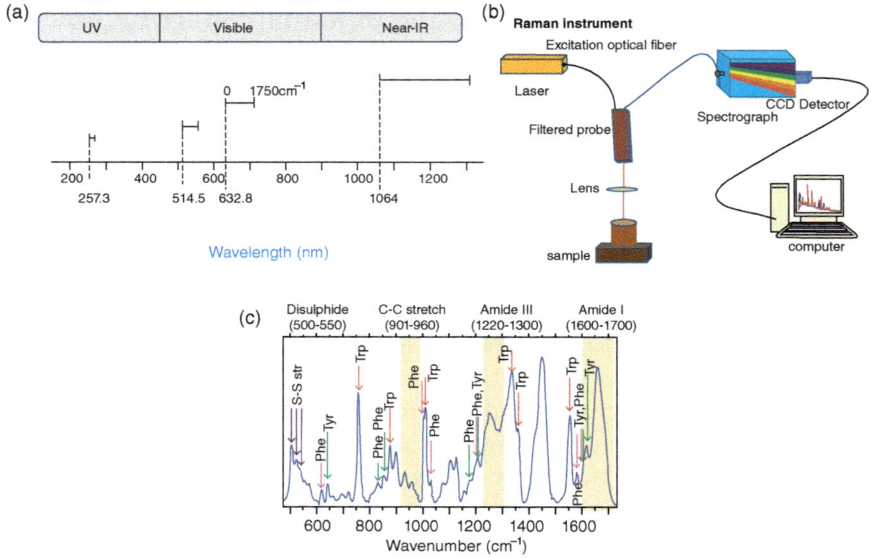

Figure 3.1 (a) The region 0–1750 cm^{-1} of the Raman scattering spectrum described (from left to right) for excitation at 257.3 nm (Ar+ laser, second harmonic of the 514.5 nm line), 514.5 nm (Ar+ laser, green), 632.8 nm (HeINe laser, red), and 1064 nm (Nd:YAG laser, near infrared). (b) Raman instrumentation. (c) A baseline-corrected Raman spectrum of hen egg white lysozyme dissolved in water. Reproduced with permission of Elsevier [6]. Creative Commons Attribution License (5315770177138).

studies of model peptides and amino acids. The interpretation is frequently based on changes in peak shape rather than on individual band assignments [7].

In general, when monochromatic light of frequency $\nu_0 o$ irradiates a molecule, the oscillating electric field E induces a small discrepancy between centers of negative (electronic) and positive (nuclear) charge, and thus an induced dipole moment is created in the molecule, which is well known by polarization. To a good approximation, the induced dipole moment P is proportional to the incident electric field strength E:

$$P = \alpha E,$$

where α represents the proportionality factor and is the molecular electric polarizability tensor. It has been noticed that the vibrations localized in the planar amide bond are expected to yield relatively high Raman intensities because of the large polarizability changes associated with in-plane stretching vibrations involving peptide C=O and C—N bonds. In-plane vibrations of the rings of aromatic side chains (Trp, Phe, Tyr) are also expected to be intense in the Raman spectrum. Stretching vibrations of C—C, C—N, and C—O bonds, especially concerted or symmetrical displacements of side-chain skeletons or carboxylates, lead also to intense Raman bands. In contrast, it has been documented that the exocyclic substituents have weak Raman effects. Bending and stretching modes of C—H, N—H, and O—H bonds are generally weak in the Raman spectrum. However, the collective Raman intensities resulting from the large numbers of such weak groups in a protein can be high. Intense Raman bands are associated with vibrations of groups that involve displacements of heavy atoms, such as sulfur. Thus, relatively intense Raman bands are associated with C—S stretching modes of Met and Cys, S—S stretching of cystine, S—H stretching of cysteine, and Zn—S stretching in zinc metalloproteins [8].

Polarization is a unique and useful characteristic of Raman scattered light. Although the incident laser beam is plane-polarized, it is calculated that the Raman scattered light is not generally polarized in the same plane. The change of the plane of polarization upon scattering is dealt quantitatively by defining the depolarization ratio ρ ($\equiv I \perp/_{II} \equiv I\perp / I_{II}$), which is the ratio of intensities of light scattered along directions perpendicular ($I\perp$) and parallel (I_{II}) to the plane of the incident light. In isotropic ensembles of molecules (solutions), measurements of ρ are helpful in distinguishing normal modes of vibration that are totally or locally symmetric (i.e. which retain the symmetry of the molecule or subgroup and for which $\rho < 0.75$) from those modes that are non-totally symmetric ($\rho = 0.75$). In anisotropic ensembles of molecules (single crystals, oriented membrane multilayers, or oriented fibers), the Raman polarization can yield information on the orientations of molecules or their subgroups [9].

3.2.2 Instrumentation

The basic instrumentation for a Raman spectrum consists of a monochromatic light source (laser) for excitation of the Raman scattering, a diffraction grating spectrometer for analysis of the scattered frequencies, and a detector for capturing the scattered photons (for better understanding see Figure 3.1b). Wavelengths in the range of near-ultraviolet (= 200 nm) extending to the near infrared (= 1000 nm), generally defined as laser wavelengths, have been used. And the selection of wavelength depends on the type of sample and extent of the experiment. Typical excitation wavelengths include 488.0 and 514.5 nm of argon and 632.8 nm of helium-neon lasers. Intracavity frequency-doubled argon lasers, which provide continuous-wave (CW) laser emission in the ultraviolet (264, 257, 248, 244, 239, and 229 nm), have become available recently for ultraviolet resonance Raman (UVRR) spectroscopy of proteins that absorb in the ultraviolet. In the UVRR spectra of such proteins, the spectrum is dominated by the resonance-enhanced Raman bands of the aromatic amino acid side chains.

The detector in Raman spectroscopy is gradually evolving from the single-channel photomultiplier and multichannel diode array technologies to the more efficient charge-coupled device (CCD) detector. In recent years, the liquid-nitrogen-cooled CCD camera, which virtually eliminates dark current noise and permits rapid multichannel detection, has come into wide use for the detection of Raman scattered photons [10].

In Raman spectroscopy, the vibrational mode has been measured as it is very sensitive to the chemical composition and strength of bonds. Due to their large size, the vibrational spectra of proteins are very complex. Thus, a couple of vibrational modes can be used to analyze the peptide structure (Figure 3.1c).

3.3 Advances in Raman Spectroscopy Techniques

Apart from classical Raman spectroscopic studies, new advancements have been made in Raman spectroscopy in combination with other techniques to obtain more insights into the determination of the structure of proteins. Some of these new areas are described in the following sections.

3.3.1 Resonance Raman Spectroscopy for Protein Analysis

Resonance Raman spectroscopy (RRS) is an extension of standard Raman spectroscopy with very high sensitivity for compounds containing chromophores in a complex mixture at low concentrations [11]. In Raman, the light of any wavelength (visible, near-IR, near-UV, X-rays) can interact with molecules one wishes to analyze, leading to inelastic and nonspecific scattering. If the incident photon with frequency ν interacts with molecules at the ground state (E_o), which undergoes excitation to the virtual stage (nonexistent state) and returns without any energy loss, that is called Rayleigh scattering. However, if molecules return to E_o, $\nu = 1$, excited vibrational state, the phenomenon is called stokes scattering, which occurs with loss of frequency ($\nu' = \nu - \nu_\nu$) (Figure 3.2). In anti-Stokes scattering, a molecule already present in an excited vibrational state (E_0, $\nu = 1$) and after interaction with an incident photon returns to a lower vibrational state (E_0, $\nu = 0$) with addition of frequency ($\nu'' = \nu + \nu_\nu$). Unlike other methods, in RRS, the excitation of molecules from the ground electronic state (E_0, $\nu = 0$) to the virtual state, which is very much close to the first excited electronic state (E_1, $\nu = 1$), occurs when the wavelength of incident photons is in resonance with the absorbance band. In the standard Raman scattering or off-resonance Raman scattering, the wavelength of incident light does not resonate with the absorbance band of the analyte. Unlike in standard Raman scattering, in RRS the exciting wavelength lies within the absorption band. The off-resonance Raman scattering spectrum comprises all the vibrations occurring within the molecule, while in RRS a small set of vibrations dominates the spectrum because of the use of excitation laser that has a particular wavelength in the vicinity of electronic transitions

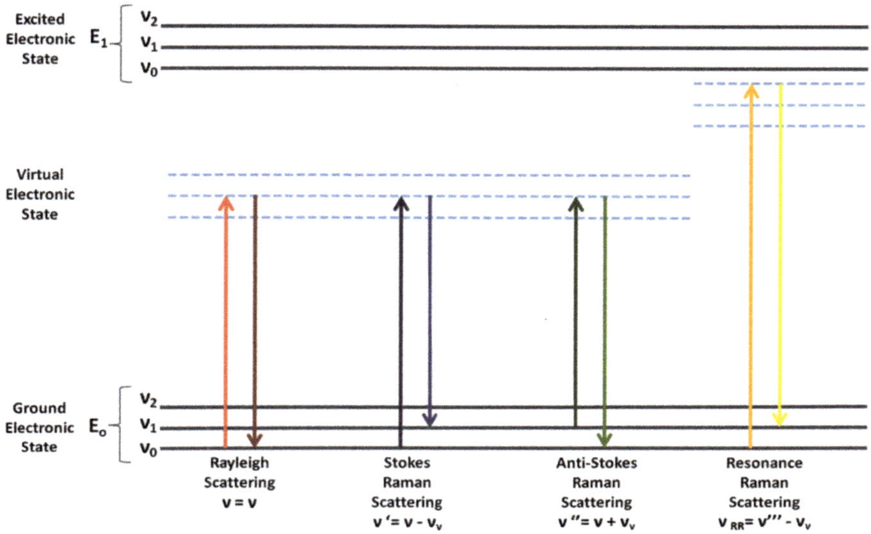

Figure 3.2 Representation of the Raman scattering principle.

associated with a particular chromophore. The RRS intensity of particular band for a system can be enhanced up to the factor of 10^8 [12]. These special features of RRS allow it to study proteins sensitively and selectively [13]. RRS can be classified into ultraviolet resonance Raman spectroscopy (UVRRS) and visible Raman spectroscopy (VRRS) [14]. UVRRS comprises wavelengths in the range of 193–240 nm extensively used for investigating the secondary structure of proteins. However, VRRS is widely employed for colored compounds to study chromophore behavior, such as the heme group in hemoglobin.

3.3.1.1 Ultraviolet Resonance Raman Spectroscopy

Ultraviolet resonance Raman spectroscopy (UVRRS) is widely employed to abstract spectral information from vibrations of the amide group, proline, and side chains of aromatic amino acids in protein analysis [15]. The UVRRS analysis provides precise information about the hydrogen bonding, protonation state, side-chain conformation, and secondary structure of proteins [16]. The resonance Raman enhancement effect in UVRRS provides a remarkable potential to study the dilute protein solutions by overcoming the disadvantages of conventional infrared spectroscopy and visible Raman spectroscopy [17]. In addition, the ability to modulate excitation wavelength leads to the enhancement of vibrational spectra for particular chromophores [18]. For example, myoglobin (Mb) analysis under the UVRRS showed the different transitions corresponding to different groups depending on the excitation wavelength [19]. UVRRS excitation of Mb at

229 nm gives the dominating spectra corresponding to aromatic rings of Trp and Tyr. In contrast, this excitation at 415 nm leads to intense UVRR spectra due to the heme ring [20]. In the case of deeper UV, excitation at 206.5 nm generate the vibrational spectra corresponding to amide bonds. Thus, UVRRS allows studying separate chromophores within the same protein by tuning the exciting wavelengths.

UVRRS can directly provide information about the secondary structure of the protein with the implementation of a quantitative methodology [21]. The proteins with known X-ray structures have been studied, and basic spectral information for α-helix, β-sheet, and unordered secondary structures have been abstracted utilizing the excitation transition at 206.5 nm. For the α-helix structure, the Raman spectrum shows the amide I, II, and III bands at 1647, 1545, and 1299 cm^{-1}, respectively, without any Cα–H bending band. For the β-sheet structure, bands at 1654, 1551, and 1235 cm^{-1} appear with Cα–H bending band at 1386 cm^{-1}. Similarly, the disordered secondary structure shows the amide I, II, and III bands at 1665, 1560, and 1267 cm^{-1}, respectively, along with Cα–H bending at 1386 cm^{-1}. The linear fitting of these primary spectrum data to the obtained UVRR spectra can generate information about the secondary structure composition. Deep UVRRS can eliminate the possibility of any interference from fluorescence based on the chromophores in proteins that have their excitation below 260 nm [22]. UVRRS also provides kinetic information for fast biological processes [23]. Depending on these, UVRRS is a unique method to investigate protein dynamics and protein structure.

3.3.1.2 Time-resolved Resonance Raman Spectroscopy

Time-resolved resonance Raman spectroscopy (TRRRS) is a powerful method to elucidate excitation species' conformational, kinetic, and structural data in a short period of time [24]. In TRRRS, a sample is investigated using a single-pulse/double-pulse CCD detector, a monochromator, and a probe laser coupled with a pump [25]. The principle of TRRRS is based on the photoreaction triggered by a pump laser beam followed by the creation of Raman scattering spectra by a second laser pulse (Raman pulse). Two types of configurations can be used in the TRRRS technique: single pulse and double pulse. In single pulse, the same laser beam is used for excitation of species in the pump laser and Raman pulse, whereas in double pulse, two different beams are used. The mechanism of molecular cooperativity of hemoglobin (Hb) can be studied using TRRRS by elucidating the Raman bands for reduced and oxidized forms [25]. In the deoxy Hb, a characteristic Raman band for oxidation state was observed at 1357 cm^{-1} corresponding to 5 coordinated iron (5C) in high spin. The 5C iron (out of heme plane) can be differentiated from 6 coordinate iron (6C, in heme plane) based on vibrational frequencies of about 10 cm^{-1} lower than 6C iron. Upon binding of oxygen (oxyHb) the Raman band shifted to 1373 cm^{-1}, which is higher than normal 6C iron. The transfer of electrons from the hem group to oxygen is responsible for this upshift, which only confirms the oxygen binding in the ferrous state. Similarly, carbon monoxide dissociation and rebinding in myoglobin (Mb) can be studied by comparing the TRRRS spectra of MbCO and deoxyMb [26].

The only disadvantage of this technique is that the laser light used for excitation can damage the sample, which can be solved by either agitation of the sample or using flow methods.

3.3.2 Surface-enhanced Raman Spectroscopy (SERS)

Fleischmann was the first to report the surface-enhanced Raman Spectroscopy (SERS) phenomenon of pyridine in 1974 on an electrochemically roughened silver electrode [27]. Later on, in 1977, Jeanmaire and Van Duyne [28] and, independently, Albrecht and Creighton [29] deduced that the rough silver electrode generates a Raman spectrum that is 10^6 times more powerful than what was predicted. Thereafter significant attention has been paid to the SERS phenomenon of several molecules.

The need for SERS arises because the primary constraint of Raman scattering is that it is a fragile phenomenon. Around 1 in 10^7 photons undergo Raman scattering. Due to the inherent poor scattering process, a low Raman signal is observed. Generally, a concentration as high as 10 mg/ml of protein is required to obtain good Raman spectra, which may not be economical. Besides, sufficient a concentration may not be available in clinical samples or field conditions. Thus, efforts are being made to boost the Raman signals of proteins.

SERS is a type of Raman spectroscopy in which an analyte molecule on getting adsorbed to a metallic surface having nanoscale roughness achieves a remarkably increased Raman scattering [31]. SERS integrates interactions between light and metal, i.e. plasmonic processes, (Figure 3.3b) along with light and molecule interactions, i.e. vibrational spectroscopy (Figure 3.3a). In order to comprehend the phenomenon, it is necessary to comprehend the two interactions explicitly.

When the incident laser impinges on the metal and dielectric interfaces, the electromagnetic wave can cause the delocalized conduction electrons of the metal to oscillate collectively. Surface plasmon resonance (SPR) occurs when the frequency of the incident light matches the inherent oscillation frequency of free electrons in the metal. The resonance frequency is affected by several factors such as particle size, shape, dielectric environment, electron density, and effective electron mass. SPR can be localized to a specific position in metal nanostructures, known as localized surface plasmon resonance (LSPR). The nanoparticles that can produce a strong LSPR effect are plasmonic nanoparticles (PNP), and they are typically Ag, Au, and Cu because they exhibit powerful SPR in the visible to near-infrared range. The incident light will be resonantly absorbed or scattered

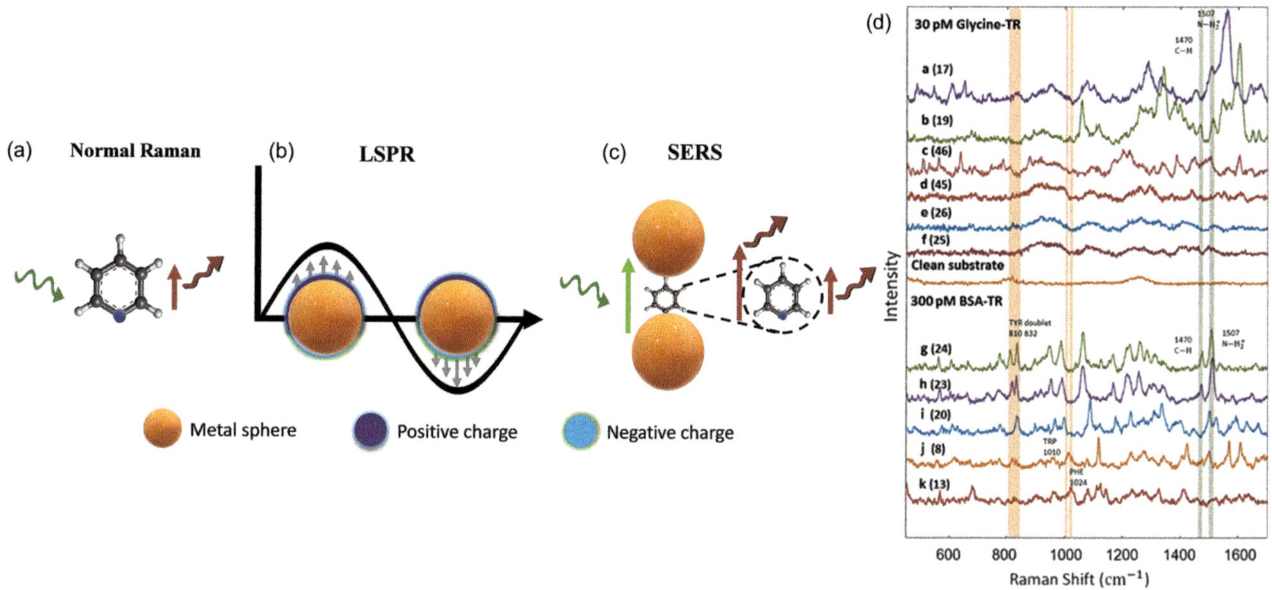

Figure 3.3 Schematics of (a) normal Raman, (b) LSPR: When irradiated with light, conductive metal electrons are excited into collective oscillations, generating an electromagnetic field highly localized in the metal-dielectric interface; (c) electromagnetic enhancement mechanism, leading to collective excitation of the Raman polarizability (red arrow) from the local electromagnetic field (green arrow) and generating the enhanced Raman signal of the molecule (red arrow), and (d) Glycine-TR and BSA-TR representative spectra gathered from specific spots on a respective substrate Copyright © 2019, Lamyaa M. Almehmadi et al. [30].

due to LSPR. As a result, incident light energy can be effectively coupled into the metal nanoparticles, resulting in a 2 to 5 orders of magnitude increase in the local electromagnetic field intensity at the nanoparticle surface, leading to a massive increase in SERS (Figure 3c).

SERS is a strong and convenient analytical tool for the structural characterization of materials that can significantly improve the poor Raman signals of molecules and materials. It is observed that the Raman signal of molecules adsorbed on the metallic nanoscale particles is enhanced by a factor of 10^2–10^{14} [32]. Hence, SERS can be utilized for sensitive and selective molecular identification [33, 34]. When combined with resonant Raman (generally referred to as surface-enhanced resonant Raman spectroscopy [SERRS]), SERS can even identify molecules at the single molecular level [35].

Using SERS, Xu et al. detected the Raman spectra of single hemoglobin (Hb) molecules [36]. The Hb molecule to Ag particle ratio was kept to 1:3 to ensure that the Raman signals come from a single molecule. At such low concentrations, only a small number of Hb molecules adhere to hotspots that can be approximated to a single molecule. Han et al. reported a reproducible Raman spectrum of lysozyme, ribonuclease B, avidin, catalase, and hemoglobin, wherein detection limit as low as 50 μg/ml for lysozyme and 50 ng/ml for catalase was achieved. The detection protocol for label-free proteins involved aggregating protein with silver colloids and acidified sulfate. This study helped in inducing strong SERS signals [37].

Because SERS can detect even single molecules, attempts are being made to develop the SERS-based immunoassay (SERBIA) for protein quantification and imaging. Shin et al. created a SERS substrate based on gold nanoparticles for protein quantification at a sub-picomolar concentration [38]. A biological sample containing a target protein (antigen) and other biological metabolites was functionalized by gold nanoparticle adsorption. Repetitive centrifugation cycles promote a sufficiently conductive coating of gold nanoparticles on the target protein's surface, allowing for effective SERS signals. A glass substrate was coated with an antibody in their experimental setup to selectively adsorb and detect the target proteins by SERS. The sensitivity of detection was comparable to the enzyme-linked immunoassay (ELISA). Hao Ma et al. investigated spacer molecules between proteins and SERS substrates and optimized them for biocompatible protein immobilization and Raman scattering enhancement. They created iminodiacetic acid-functionalized silver substrates, which are used to capture His-tagged proteins via nickel imidazole coordination. Six polypeptides demonstrated excellent SERS spectral reproducibility due to controlled immobilization. In addition, the interactions of two model proteins, the C-terminal domain of flavine adenine dinucleotide-dependent mitochondrial cytochrome c reductase Erv1 and alpha-fetoprotein, and their ligands cytochrome c and all-trans-retinoic acid, were investigated [39].

L. M. Almehmadi and co-workers created a two-step process for protein detection at the single-molecule level using SERS for medical diagnostics. In the first step, a protein molecule, i.e. bovine serum albumin (BSA) was bound to a linker, i.e. Traut's reagent (TR) in bulk solution. In the second step, a protein-bound linker was crosslinked with the gold surface by the sulfhydryl group. Glycine bound TR was utilized as a control sample to determine the contribution of protein in the SERS spectra. The electrochemical deposition was used to make gold SERS substrates. BSA-TR and GR-TR adducts were attached to the SERS substrate using extremely low concentration solutions. The samples exhibited typical single-molecule SERS behavior, such as spectral fluctuations, blinking, and Raman signal generation (Figure 3d). Principle component analysis was used to examine the fluctuating SER spectra. They found that the protein moiety contributed significantly to the spectra, indicating that the method can detect a single protein molecule. The developed methodology has the potential to serve as a new platform for medical diagnostics [30].

The importance of the SERS technique is not only that it amplifies the Raman signal but also gives high-resolution spectra in samples containing water and reduces the fluorescence background that usually interrupts the Raman signals of biological molecules [40]. While the SERS technique has several advantages and has much potential as a novel methodology in detection, it is yet not a fully developed technique in several applications. The aggregation phenomenon of metallic nanoscale surfaces cannot be controlled [41]. Also, it is challenging to produce good quality and reproducible nanosurfaces, which are essential for SERS detection, and thus it needs further study.

3.3.3 Tip-enhanced Raman Spectroscopy

Tip-enhanced Raman scattering (TERS) was invented by Wessel in 1985. Zenobi, Kawata, Anderson, and Pettinger independently reported TERS results for the first time in 2000, demonstrating the feasibility of TERS [42–44]. TERS combines the benefits of surface-enhanced Raman spectroscopy and scanning probe microscopy [45]. TERS, unlike other electron spectroscopy and microscopy techniques such as scanning electron microscopy, transmission electron microscopy, and X-ray photoelectron spectroscopy, can be used in an ambient environment and is well suited for investigating samples in aqueous media [46]. Even though super-resolution fluorescence microscopy techniques can be used to image biological samples below the diffraction limit, the fluorescent labels make it impossible to observe the samples in their natural state [47]. TERS, as a label-free technique, can be used to study molecules directly in biological samples to investigate their chemical composition and molecular dynamics. With the addition of a microscope to Raman spectrometers, it is possible to image samples while reading their Raman signature. On the other hand, microscopy analysis is limited to diffraction-limited resolution and low signals due to the weak Raman cross-section. SERS has paved the way for signals to be detected at extremely low concentrations. Tip-enhanced Raman spectroscopy (TERS) can be used to improve spatial resolution [48]. TERS combines a Raman spectrometer with an atomic force microscope (AFM) or scanning tunnelling microscope (STM) (Figure 3.4a). Using a sharp metal tip, the spatial resolution can be significantly increased, while maintaining the enhancement factor for detecting a single molecule. TERS has piqued the interest of scientists all over the world since its discovery.

A metallic tip (typically made of gold or silver) serves as a nano-light source to scan the sample. An excitation laser illuminates the tip. Surface plasmons and localized surface plasmon resonance enhance the intensity of the excitation light at the tip's apex. The tip functions as an illuminating antenna, increasing Raman

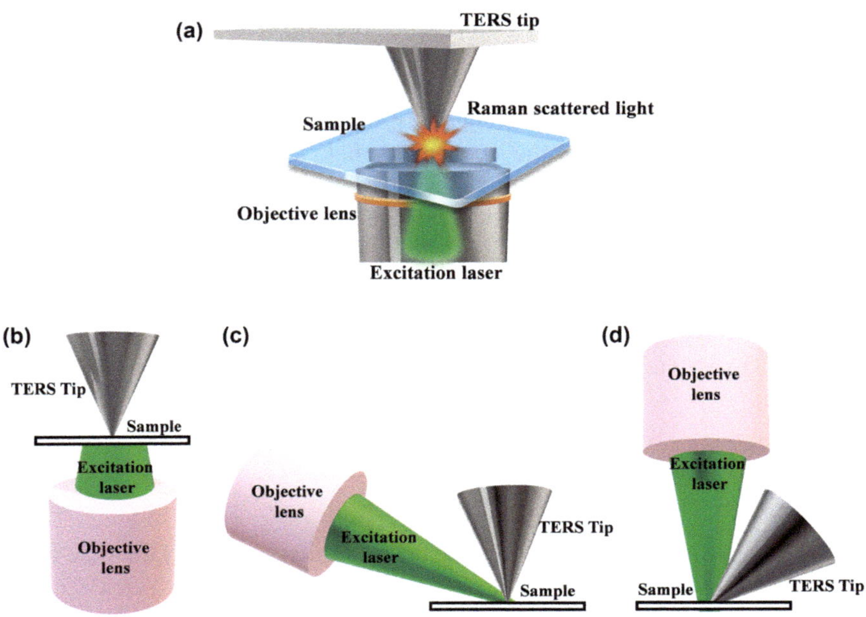

Figure 3.4 (a) TERS setup is depicted schematically. A schematic diagram illustrating the AFM-TERS principle in transmission mode. The localized surface plasmon resonance at the tip-apex between the tip-localized surface plasmon and laser-photons enhances and confines the electromagnetic field, resulting in an increase in Raman signal from analyte molecules in close proximity to the tip-apex. (b–d) General figures of most frequently used TERS configurations: (b) bottom, (c) side, and (d) top illumination setup.

sensitivity by a factor of 10^3–10^7 [49]. In TERS, the sample volume probed with the tip apex is much smaller than in SERS. In SERS and TERS, sensitivity and resolution are primarily interpreted in terms of electromagnetic enhancement mechanisms [50], though signal enhancement caused by the chemical interaction of a TERS tip with the sample has also been reported [51]. When the local surface plasmon wavelength of the tip apex matches the wavelength of the excitation laser, an even more localized electrical field enhancement is observed. The local surface plasmon wavelength of the TERS tip can be varied by changing the material, radius, or roughness of the tip apex, which can then be aligned with the wavelength of the excitation laser. The intensity of the observed electrical field at the tip apex is several orders of magnitude greater at resonance than at off-resonance. Although local surface plasmon resonance (SPR) has been witnessed for many metal nanoparticles, only silver and gold are visible-light resonant with lasers. The SPR wavelength of a silver TERS tip is in the blue and green light region, whereas that of a gold TERS tip is in the yellow and red light region. However, it is also affected by the radius of the tip apex. For example, the wavelength of a silvertip with a radius of 10 nm may be in the ultraviolet (UV) region. Owing to the reduced penetration depth of UV photons, such tips could be used to conduct UV-TERS with a lesser background signal from the substrate [52]. The intensity of the Raman signal from the analyte molecules is proportional to the fourth power of the local electric field near the TERS tip [50]. As a result, the enhancement of Raman signal (ρ) for near-field and far-field electric-field intensity E_{NF} and E_{FF}, respectively, is equivalent to

$$\rho = \left(\frac{E_{NF}}{E_{FF}}\right)^4.$$

In a TERS experiment, the Raman enhancement factor (EF) is calculated using [53],

$$EF = \left(\frac{I_{Tip-in}}{I_{Tip-out}} - 1\right)\frac{A_{FF}}{A_{NF}},$$

where $I_{Tip\text{-}in}$ and $I_{Tip\text{-}out}$ are the Raman peak intensities measured with the tip in contact with the sample and retracted from it, respectively; A_{FF} is the far-field laser probe's area; and A_{NF} is the effective area of a TERS probe, which is usually calculated from the tip apex's diameter. The term $\left(\frac{I_{Tip-in}}{I_{Tip-out}} - 1\right)$ is also known as "contrast."

Optical geometries of a TERS setup can be categorized into three types: bottom (transmission mode), side, and top illumination. Reflection modes are also used to describe side and top illumination. In transmission mode (Figure 3.4a), a laser beam is focused onto the sample through a transparent substrate by an objective lens having a larger numerical aperture, typically greater than 1.0 [53]. In this geometry, a radially polarized beam has a higher focus and a greater longitudinal electric field along the tip shaft than a linearly polarized beam in the focal place. Theoretical calculations and experimental results show that electromagnetic enhancement using a radially polarized beam can be more than four times stronger than using a linearly polarized beam [54]. Although the engineering execution of the transmission mode setup is simpler, it is constrained by the mandate of a transparent substrate. A linearly polarized beam (p-polarization) [55] is focused onto the tip apex from the side of the tip using a long working distance objective lens in the side illumination setup (Figure 3.4b). Since the polarization and tip direction are the same, a powerful electromagnetic enhancement at the tip apex could be acquired even with a low numerical aperture objective lens. This technique is beneficial for TERS characterization of nanoelectronic devices, where substrates are typically opaque.

In the top illumination setup, the laser is concentrated from the top onto the end of an inclined STM tip [56] or a nose-type AFM tip [57] (Figure 3.4c). Zhang et al. reported another type of reflection mode setup wherein the laser could be focused over the apex of the mirror [58]. In this setup, a parabolic mirror was used to concentrate the laser over the apex of a TERS tip normal to the analyte [58]. This geometry has the benefit of being free of chromatic aberration, having a high numerical aperture (≈1), and having a tight laser focus.

Krasnoslobodtsev et al. used AFM and TERS to investigate the structure of amyloid fibrils formed by the CGNNQQNY peptide from the yeast prion protein Sup35 [59]. It was discovered that there are two types of amyloid fibers with different morphologies that form at pH 5.6 and 2.0. using TERS. By analyzing the amide I and III bands, it was discovered that the secondary structure of fibrils at pH 5.6 was a mixture of β-sheet, random coil, and α-helix, whereas that formed in pH 2.0 solution were primarily composed of β-sheets. Using an empirical equation below, the dihedral angle Ψ of the backbone was calculated using the peak position of the amide III band under non-resonant conditions.

$$\gamma^{\mathrm{beta}}{}_{\mathrm{AmideIII}} = 1239\left(\mathrm{cm}^{-1}\right) + 54\left(\mathrm{cm}^{-1}\right) \times \sin\left(\Psi + 260^{\circ}\right).$$

For fibrils at pH 2.0, a distinct amide III band at 1223 cm^{-1} was observed, which corresponds to a Ψ angle of 135°, indicating the presence of parallel β-sheets in fibrils. In contrast, at pH 5.6, a band at 1237 cm^{-1} was observed corresponding to a Ψ angle of 155°, indicating the presence of antiparallel β-sheets [59]. Several other studies [60, 61] used TERS to perform detailed protein analyses.

3.3.4 Polarized Raman Spectroscopy

Among the recent advancements over traditional Raman spectroscopy is the polarized Raman spectroscopy (PRS) [62] (Figure 3.5a). It is measured by

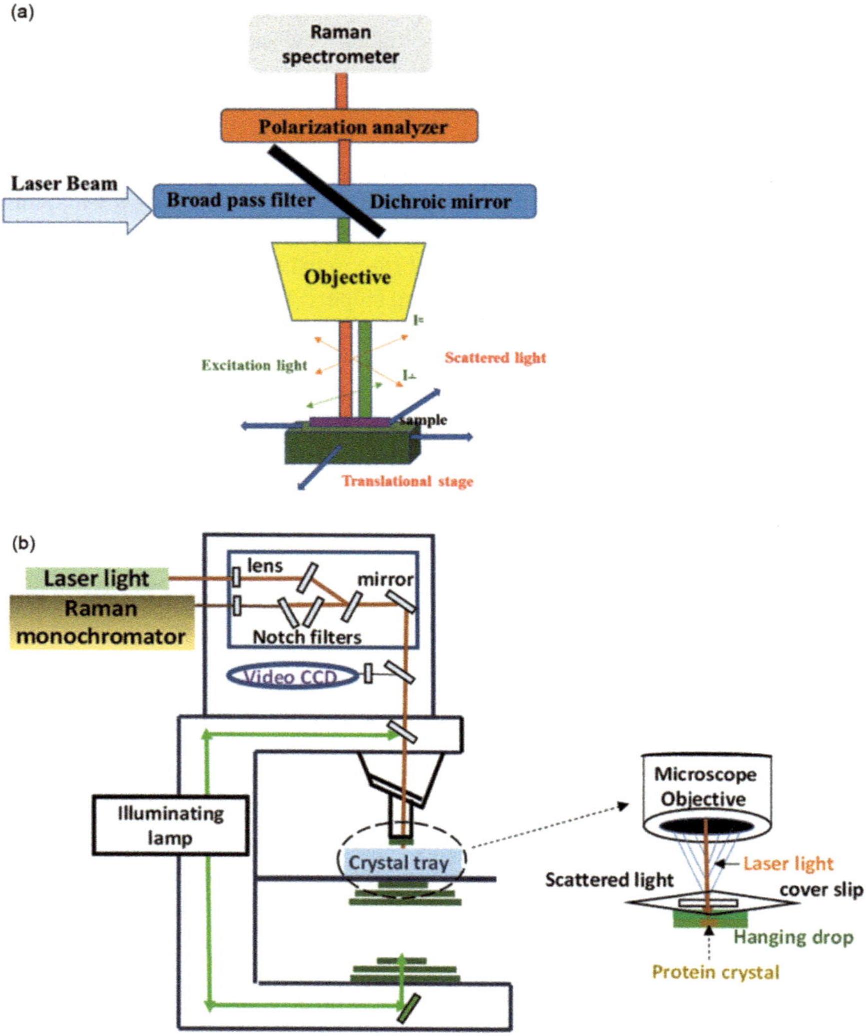

Figure 3.5 (a) The configuration of the polarized Raman imaging (PRI) with alpha 300 RSA+ (WITec, Ulm, Germany) [63]/with permission of Elsevier, and (b) basic technique and instrument used in Raman crystallography [64], Carrey et al. (2006) Annual Reviews.

acquiring both the polarization parallel and perpendicular to the excitation light using polarizing filters. PRS is based on the fact that the orientation of the molecular symmetry axes with respect to the polarization of the incident laser beam can greatly influence the efficiency of Raman scattering. PRS is useful for the assignment of vibrational modes, identification of symmetry of bond vibrations, determination of molecular orientations and conformation of the molecules in isotropic and ordered phases, the characterization of molecular or solid-state structure, single differentiation crystals vs polycrystalline materials, and detection and characterization of disorder, both in the crystalline and amorphous phases [63]. All molecules are more or less symmetric, and the excitation of the molecule on one or another symmetry axes can lead to a different intensity of Raman scattering.

In addition to determining the structure of proteins, this technique also serves as a sensitive and selective fingerprint of three-dimensional structure, intermolecular interactions, and dynamics. Recently, more modifications and advancements have been done to find the Raman spectroscopy with polarized properties to determine the structure of large supramolecular assemblies, such as detailed structural parameters in viruses and bacteria. Recently, Raman tensors of localized base and backbone vibrations of double-helical d(CGCGCG) in an oriented single crystal and later, α and β conformations of double-helical DNA and the α conformation of double-helical RNA were determined [65]. In another instance, using tensors for key Raman markers of tryptophan [66], the precise orientation of the side chain of Trp26 in the coat protein subunit of the viruses of the native assembly has been determined. Masic et al. [67] used the PRS to characterize the anisotropic response of the amide I band of collagen as a basis for evaluating three-dimensional collagen fibril orientation in tissues.

3.3.5 Raman Crystallography

Recently, Raman spectroscopic and X-ray crystals have been combined in a way to find insight into the protein structure (Figure 3.5b). Soon after discovering "bio Raman spectroscopy" in R. C. Lord's laboratory in the 1960s, it was realized that Raman spectroscopy could be used to compare the conformations of proteins in solution with those in single crystals. Small differences observed between crystal and solution Raman data were interpreted as changes in amino acid side-chain conformation. Similarly, resonance Raman spectroscopy was used to study the porphyrin moieties in cytochrome P450 and myoglobin protein crystals. The active site porphyrin structures were found to be identical for the crystal and solution phases. In the 1980s and 1990s, the development of Raman microscopy took place, where an optical microscope is coupled to a Raman spectrometer to collect Raman signals from sample regions as small as a few micrometers [68]. Thus, Raman crystallography depicts a clear picture of molecular detail of the events occurring in protein crystals and other protein-based bodies. In Raman crystallography, a Raman microscope is used to study chemical changes within a single crystal of macromolecules combined with X-ray crystallographic studies. A commercially available crystallization tray is mounted on the microscope stage, and a crystal is placed in hanging drop geometry on top of the well. The crystal is monitored using a long focal length objective. The crystal and the focusing spot are in the order of a few micrometers and hence cannot be viewed via naked eye. Usually, a microscope is operated in the non-confocal mode with a focal volume of 20 mm in diameter and 30 mm in depth to achieve maximum light throughput. A minimal dimension of 30 mm of the crystal is ideal for optimal spectral quality. Carey's group investigated the binding of methyl-malonyl-CoA (MM-CoA) to the 12S subunit of the enzyme transcarboxylase. The MM-CoA has a strong peak at 724 cm^{-1} that is used to follow the interaction of MM-CoA with transcarboxylase. It is observed that the intensity of the band increases with an increase in time, which can be due to the binding of MM-CoA with 12S. Carey's group used Raman crystallography to differentiate the crystal structure of different insulin mutants. For example, insulin exists as a hexamer. However, the mutant's insulin may exist as a dimer or trimer. The full width at half maximum (FWHM) of amide I is a good measure of the heterogeneity of crystal structure. It is reported that the FWHM of amide I profile progressively narrows, going from a monomer (54 cm^{-1}) to a dimer (48 cm^{-1}) to a hexamer in solution (42 cm^{-1}) and finally to a hexamer in the crystalline phase (37 cm^{-1}). The line broadening can be attributed to rapidly interconverting conformational states in mutants and represents heterogeneity [68]. Single protein crystals provide an impressive example of correct protein folding and three-dimensional array formation. However, Raman microscopy shows that during some ligand soaking-in experiments, proteins within crystals can undergo massive and unexpected conformational changes and assume largely β-sheet secondary structure, mimicking the endpoint of protein misfolded entities found in disease states. Raman microscopy is also used to provide novel information on intermediates formed on the misfolding pathway of the protein α-synuclein that lead to aggregates that play a major role in the progression of Parkinson's disease. A key aspect of the latter studies involves synergy between two techniques. Atomic force microscopy defines the shape and size of α-synuclein aggregates in vitro, whereas Raman microscopy on the same sample probes protein secondary structure [64].

In another illustration, enzyme structure has been recorded using Raman spectrum by means of a single crystal in a hanging drop placed inside a well of a standard crystallization tray standing on the microscope stage. In an application of Raman crystallography, the reaction of RNA polymerase has been recorded. The RNA polymerase (RNAP) reaction is highly complex; however, it is made tractable using Raman crystallography by recording the isotopically labelled substrates and the RNAP reaction in crystals in a slow time lapse. Thus, this technique provides a new opportunity to capture the unstable intermediates on the reaction pathway by spectroscopy and crystallography [69] (Figure 3.6).

Raman Crystallography also finds applications in cell biology [70] and heme-protein crystal studies. Various other reports also document the single-crystal

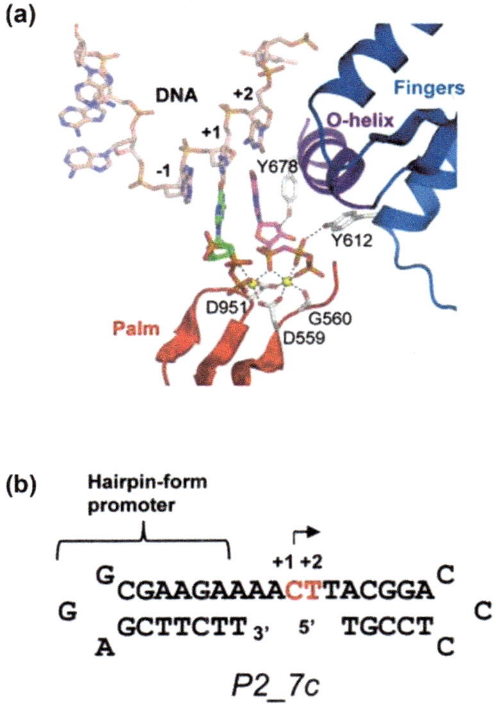

Figure 3.6 RNAP active structure and promoter DNA sequence. (a) The active site structure of the transcript initiation complex (SCII), including the N4 mini-vRNAP (fingers, palm and ohelix are labelled with colors), promoter DNA (pink tube), two GTPs (ball and stick model, +1 and +2 are green and magenta, respectively), and two divalent metals (yellow balls). (b) Sequence and secondary structure of the promoter DNA constructs used for Raman crystallography experiments. Nucleotide-binding sites (+1 and +2) for the transcript initiation complex formation are colored in red [69]. Reprinted with permission from Michael L. Gleghorn et al. Copyright © 2011, American Chemical Society [69]; Chen, Y., Basu, R., Gleghorn, M. L., K. S., Carey, P. R. "Time resolved events on the reaction pathway of transcript initiation by a single subunit RNA polymerase: Raman crystallographic evidence," *J Am Chem Soc.* 2011 August 17; 133(32): 12544–12555. doi:10.1021/ja201557w.

(Figure 3.7) study of the protein to determine the structure orientation using Raman crystallography [71–73]. Crystals with lower solvent content should yield better spectra, at least in non-resonant mode. When selecting appropriate cryo-solutions, care should be taken since some common chemicals (e.g. ammonium sulfate, polyethylene glycol, and glycerol) give extreme Raman bands. When 785 nm light is used, ice is sufficiently transparent so that cryoprotection is generally not required; this eliminates potentially contaminating bands. When excitation

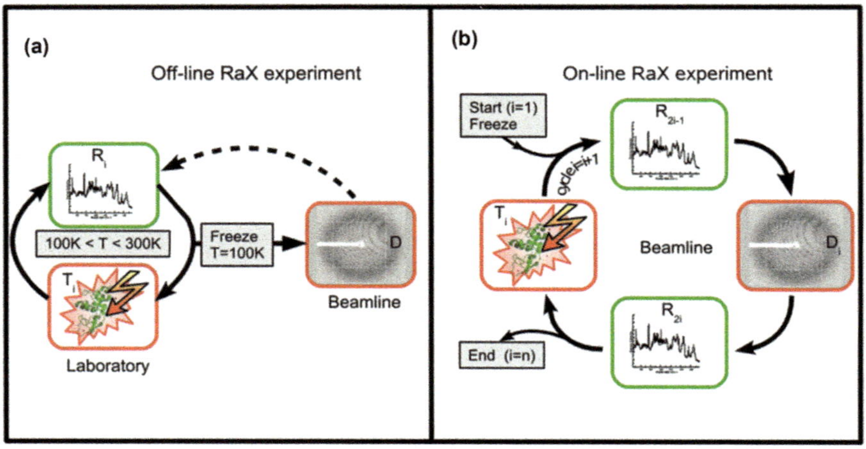

Figure 3.7 On-line and off-line experimental procedures. (a) The typical off-line mode monitors changes triggered (T) in the crystal at a suitable temperature (100–300 K) by Raman spectroscopy (R) prior to cooling (or flash-cool) the crystal down to 100 K to trap the desired state for MX diffraction (D). (b) Several variations of the on-line mode are possible where continuous Raman monitoring can provide real-time information during the MX experiment. This method is ideal for collecting optimum pre- and post-MX Raman spectra since the crystal can be rotated to an identical position for each spectral acquisition. [71], McGeehan et al. (2011) / with permission from ELSEVIER.

at a shorter wavelength is performed, a cryoprotectant is required to ensure sample transparency and deep light penetration (e.g. at 633 nm and especially at 514 nm). It has been noticed that the enhanced Raman bands of interest tend to cover the cryoprotectant lines, thus providing suitable conditions for Raman X-ray crystallography (RaX) experiments.

3.3.6 2D-COS Raman Spectroscopy

Two-dimensional correlation spectroscopy (2D-COS) is a powerful technique to obtain spectral data under the influence of external perturbations on a given system [74, 75]. It provides new insights to understand the systems by overcoming the limitations of 1D spectral analysis. 2D-COS is useful for sorting out important information when applied to Raman as well as other spectroscopic techniques [76]. The external perturbation leads to the formation of dynamic spectra on the two-dimensional scale and creates two types of plots—synchronous and asynchronous [6]. The synchronous plot gives information about the types of events occurring in one phase, while the asynchronous plot provides the out-of-phase relation, which helps determine the sequence of events. The 2D-COS technique has the potential to resolve the overlapped peaks and track the changes in the sequence of dipole reorientation.

The enhancement in resolution and determination of consecutive changes in spectral intensities make the 2D-COS an effective tool to study the structural changes and kinetics in a particular protein sequence [77, 78]. The main perturbations in protein analysis includes the temperature [79], concentration of protein [80], chemical denaturants [81], pH [82], pressure [83], and H/D exchange [84] (Figure 3.8). Among these, temperature perturbations are commonly analyzed. In combination with 2D-COS, Raman spectroscopy provides new perceptions at the molecular level to investigate the protein folding/unfolding mechanism, including aggregation and fibril formation from β-type proteins.

In combination with 2D-COS, Raman spectroscopy was used to study the qualitative transition of the secondary structure of silk I to silk II via the formation of β-sheet fibrils at various pH and concentrations of Ca^{2+} ions [85]. The analysis showed the broad amide I bond conversion to four peaks at 1645, 1655, 1666, and 1685 cm^{-1}. These peaks indicated the distorted β-sheet conformation at 1685 cm^{-1}, silk II (β-sheet) conformation at 1666 cm^{-1}, silk I conformation at 1655 cm^{-1}, and intermediate formation between random coils and silk I conformation at 1645 cm^{-1}. The analysis of silk fibrils by 2D-COS Raman confirms the structural changes from β-sheet, distorted β-sheet to random coil upon pH variation 4.8 to 8.0. 2D-COS Raman was also used to elucidate the α-helix-to-β-sheet transition in poly(L-lysine) upon temperature variation from 4 to 52°C [86]. The transition proceeds intramolecularly, which further leads to fibril formation. The investigation of the transition mechanism in silk fibrils by utilizing the 2D-COS Raman and 2D-NMR correlation techniques can be carried out [87]. The Fermi doublet band (850 and 830 cm^{-1}) arises due to the hydrogen bonding of the Tyr phenolic-OH in Raman and acts as an indicator for fibroin structural changes. The I_{850}/I_{830} ratio displayed that the H-bonding due to the Tyr phenolic-OH was directly proportional to K^+ ion concentration, increasing the possibility of antiparallel fibrillar β-strands existence. A Raman and CD spectroscopy combination followed by 2D-COS has been utilized to investigate the 2D and 3D structure of truncated hepatitis C virus core protein [88]. The perturbations were carried out by H-D exchange, where the proton of Arg was exchanged first, indicating the Arg on the surface of protein aggregates followed by unordered and β-sheet structure formation.

The human hair keratin fiber analysis was carried out under gradual stress using 2D-COS Raman, which showed the conformation changes from α-helix to β-sheet via the breakage of disulfide bonds [89]. The 2D-COS Raman has also been applied for the qualitative identification of microorganisms that cause gastroenteritis: *S. chroleraesuis, E. coli, and S. flexneri* depending on the difference in the amide-I band of proteins [90]. This method provides the low-cost, reagent-free, and fast identification of these bacteria compared to any other method. In a recent report, the sequence and steps involved in the bovine serum albumin (BSA)

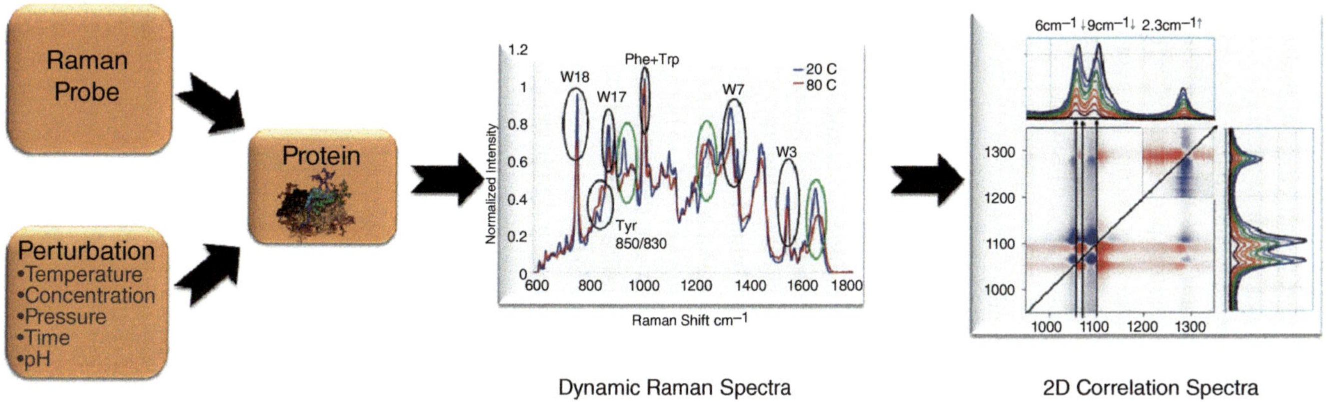

Figure 3.8 Protein analysis by Raman 2D-COS.

unfolding have been studied by combining 2D-COS Raman spectroscopy at different concentrations of guanidine hydrochloride (GuHCl) [90]. The results demonstrate that small turns that connect the helices in BSA initiate the whole protein's unfolding. The study of amide I and III regions of BSA showed that there were no significant changes in the secondary structure at lower concentration (2 M) of GuHCl because connecting segments remained stable. As the concentration of GuHCl (2–4 M) increased, the accumulation of random coils and β-sheets through the connecting segments led to the unfolding of helices and complete denaturation of BSA. Such a deep study opens a new door to study the various factors related to proteins such as structure, size, and stability of proteins and formation and structure of intermediates.

2D-COS Raman has been employed to study the pH-instigated formation mechanism of the α-lactalbumin/oleic acid (ALA/OL) complex [91]. The results confirmed the ALA/OL complex formation is accelerated by the changes in β-sheet and side chains of Tyr, Trp, and Phe residues below pH 3.0. This study also revealed that bands from the side chains of Lys, Glu, His, and Thr residues confirm their participation in the formation of the ALA/OA complex.

3.4 Applications

One of the most important applications of Raman spectroscopy in protein structure analysis is related to structural characterization and changes caused by different stages of diseases due to the folding and misfolding of the proteins [92–94]. Raman spectroscopy widely found applications for change and characterization of protein structure, thus making Raman better studied and applicable in the field of medicine to determine the causes of the disease associated with protein misfolding. Identifying specific bands associated with the secondary structure has been achieved using polypeptides and proteins with known α-helix and β-sheet elements [95–97]. Ease in handling, no interference of water, and a very small quantity of samples makes it an attractive choice for researchers analyzing complex molecules such as proteins or peptides.

Vibrational spectra of the free amino acid in an aqueous solution are generally used to decode the complexity of the protein spectra [98–100]. Many groups have documented detailed assignments of amino acid vibrational bands based on ab initio calculations [101]. However, very few reports concern themselves with the Raman spectra of di- or tripeptides and aim to determine how Raman bands of amino acid side chains are affected by the presence of the peptide bond [102–104]. Thus in a recent report, Raman spectra of a series of tripeptides with the basic formula GlyAAGly where the central amino acid differs were measured in H_2O [104]. Their analysis contributes to an unambiguous interpretation of the protein

Raman spectra, which is helpful in monitoring the biological reactions involving amino acid side chains alteration.

Furthermore, the application of the Raman is also extensively used in food protein products analysis. The history of foodstuff adulteration has accompanied mankind since food became a commodity for sale. The bread bakers added chalk, and brewers replaced part of malt with beet sugar during the brewing process, which might cause food adulteration. The gluten protein, which is a food protein product, was chemically changed by varying levels of sodium stearoyl lactylate (SSL), and the extent of modifications (secondary and tertiary structures) of this protein was analyzed by using Raman spectroscopy. Analysis of the amide I band showed an increase in its intensity mainly after adding 0.25% of SSL to wheat flour to produce modified gluten protein, pointing to the formation of a more ordered structure. The data show that the native conformation of the gluten protein is affected after the treatment with an SSL emulsifier. The adulteration of food and feed grains and oilseeds by fungal mycotoxins has raised worldwide concerns about animal and human health. The most commonly occurring mycotoxins in food and feed grains are aflatoxins and fumonisins, produced by the aspergillus and fusarium fungal strains. Mycotoxin contaminated grains and oilseeds are toxic and carcinogenic to humans and animals. As a result, they are not valued in the markets for food and feed and can result in economic loss for growers, handlers, and food and feed processors. In addition to this, some bacterial and viruses structures were also determined using Raman spectroscopy.

3.5 Conclusion

Raman spectroscopy has found its application to expand the horizon for a better understanding of various biophysical processes associated with proteins, such as the nature and strength of hydrogen-bonding interactions, protein-ligand interaction, and protein-folding studies. Recent developments in instrumentation have vastly improved the sensitivity and selectivity of Raman spectroscopy to study proteins. The exploration and elaborative depth study further helps understand and eradicate the root cause of the diseases associated with proteins structure.

References

1 Alberts, B., Johnson, A., Lewis, J., Raff, M., Roberts, K., and Walter, P. (2002). Analyzing protein structure and function. https://www.ncbi.nlm.nih.gov/books/NBK26820 (accessed 31 August 2020).

2 Alberts, B. (1998). The cell as a collection of protein machines: Preparing the next generation of molecular biologists. *Cell* 92: 291–294.

3 Ellis, R.J. (2006). Molecular chaperones: Assisting assembly in addition to folding. *Trends Biochem Sci* 31: 395–401.
4 Aboul-Enein, Y., Bunaciu, A. and Fleschin, S. (2014). Evaluation of the protein secondary structures using Fourier Transform Infrared Spectroscopy. *Gazi Univ J Sci* 27(1): 637–644.
5 Raman, C.V. F.R.S. Indian. (1928). A new radiation. *J Phys* 2: 387–398.
6 Kuhar, N., Sil, S., and Umapathy, S. (2021). Potential of Raman spectroscopic techniques to study proteins. *Spectrochim Acta A Mol Biomol Spectrosc* 258: 11971–11996.
7 Rostron, P. and Gerber, D. (2016). Raman spectroscopy, a review. *Int J Eng Tech Res* 6 (1): 50–64.
8 Nishimura, Y., Hirakawa, A.Y., and Tsuboi, M. (1978). Resonance Raman spectroscopy of nucleic acids. In: *Advances in Infrared and Raman Spectroscopy*, vol. 5 (ed. R.J.H. Clark and R.E. Hester), 217–275. London: Heyden.
9 Benevides, J.M., Tsuboi, M., Wang, A.H.-J., and Thomas, G.J., Jr. (1993). Local Raman tensors of double-helical DNA in the crystal: A basis for determining DNA residue orientations. *J Am Chern Soc* 115: 5351–5359.
10 Austin, J.C., Jordan, T., and Spiro, T.G. (1993). Ultraviolet resonance Raman studies of proteins and related model compounds. In: *Mvances in Spectroscopy. Vol. 20 Part A* (ed. R.J.H. Clark and R.E. Hester), 55–127. New York: Wiley.
11 Oladepo, S.A., Xiong, K., Hong, Z., Asher, S.A., Handen, J., and Lednev, I.K. (2012). UV resonance Raman investigations of peptide and protein structure and dynamics. *Chem Rev* 112 (5): 2604–2628.
12 Jakubek, R.S., Handen, J., White, S.E., Asher, S.A., and Lednev, I.K. (2018). Ultraviolet resonance Raman spectroscopic markers for protein structure and dynamics. *Trends Anal Chem* 103: 223–229.
13 Xiong, K., Punihaole, D., and Asher, S.A. (2012). UV resonance Raman spectroscopy monitors polyglutamine backbone and side chain hydrogen bonding and fibrillization. *Biochemistry* 51 (29): 5822–5830.
14 López-Peña, I., Leigh, B.S., Schlamadinger, D.E., and Kim, J.E. (2015). Insights into protein structure and dynamics by ultraviolet and visible resonance Raman spectroscopy. *Biochemistry* 54 (31): 4770–4783.
15 Austin, J.C., Jordan, T. and Spiro, T.G. (1993). Ultraviolet resonance Raman studies of proteins and related model compounds. *Advances in spectroscopy* (1986), 20: 55–127.
16 Austin, J.C., Redgers, K.R., and Spiro, T.G. (1992). Protein structure from ultraviolet resonance Raman spectroscopy. *Methods Enzymol* 226: 374–396.
17 Asher, S.A. (1993). UV resonance Raman spectroscopy for analytical, physical, and biophysical chemistry. Part 1. *Anal Chem* 65 (2): 59A–66A.
18 Palaniappan, V. and Bocian, D.F. (1994). Acid-induced transformations of myoglobin. Characterization of a new equilibrium heme-pocket intermediate. *Biochemistry* 33 (47): 14264–14274.

19 Ahmed, A., Beta, I.A., and Asher, S.A. (2005). UV–resonance Raman thermal unfolding study of Trp-cage shows that it is not a simple two-state miniprotein. *J Am Chem Soc* 127 (31): 10943–10950.

20 Chi, Z., Chen, X.G., Holtz, J.S., and Asher, S.A. (1998). UV resonance Raman-selective amide vibrational enhancement: Quantitative methodology for determining protein secondary structure. *Biochemistry* 37 (9): 2854–2864.

21 Roach, C.A., Simpson, J.V., and JiJi, R.D. (2012). Evolution of quantitative methods in protein secondary structure determination via deep-ultraviolet resonance Raman spectroscopy. *Analyst* 137: 555–562.

22 Smulevich, G. and Spiro, T.G. (1985). Nanosecond transient resonance Raman spectra of the Fe^{II} CO and Fe^{III} NO photolysis products of horseradish peroxidase. *Biochim Biophys Acta (BBA) - Protein Struct Mol Enzymol* 830 (1): 80–85.

23 Loehr, M.L. and Samnders-Loehr, J. (1993). Techniques for obtaining resonance raman spectra of metalloproteins. *Methods Enzymol* 226: 431–470.

24 Mizutani, Y. (2017). Time-resolved resonance Raman spectroscopy and application to studies on ultrafast protein dynamics. *Bull Chem Soc Jpn* 90 (12): 1344–1371.

25 Buhrke, D. and Hildebrandt, P. (2020). Probing structure and reaction dynamics of proteins using time-resolved resonance Raman spectroscopy. *Chem Rev* 120: 3577–3630.

26 Mizutani, Y. and Kitagawa, T. (2001). Ultrafast dynamics of myoglobin probed by time-resolved resonance Raman spectroscopy. *Chem Rec* 1 (3): 258–275.

27 Fleischmann, M., Hendra, P.J., and McQuillan, A.J. (1974). Raman spectra of pyridine adsorbed at a silver electrode. *Chem Phys Lett* 26 (2): 163–166.

28 Jeanmaire, D.L. and Van Duyne, R.P. (1977). Surface raman spectroelectrochemistry. *J Electroanal Chem Interfacial Electrochem* 84 (1): 1–20.

29 Albrecht, M.G. and Creighton, J.A. (1977). Anomalously intense Raman spectra of pyridine at a silver electrode. *J Am Chem Soc* 99 (15): 5215–5217.

30 Almehmadi, L.M., Curley, S.M., Tokranova, N.A., Tenenbaum, S.A., and Lednev, I.K. (2019). Surface enhanced Raman spectroscopy for single molecule protein detection. *Sci Rep* 9 (1): 1–9.

31 McCall, S.L., Platzman, P.M., and Wolff, P.A. (1980). Surface enhanced Raman scattering. *Phys Lett A* 77 (5): 381–383.

32 Stiles, P.L., Dieringer, J.A., Shah, N.C., and Van Duyne, R.P. (2008). Surface-enhanced Raman spectroscopy. *Annu Rev Anal Chem* 1 (1): 601–626. doi: 10.1146/annurev.anchem.1.031207.112814.

33 Dieringer, J.A., McFarland, A.D., Shah, N.C., Stuart, D.A., Whitney, A.V., Yonzon, C.R., Young, M.A., Zhang, X., and Van Duyne, R.P. (2006). Introductory lecture surface enhanced Raman spectroscopy: new materials, concepts, characterization tools, and applications. *Faraday Discuss* 132: 9–26.

34 Ding, S.-Y., Yi, J., Li, J.-F., Ren, B., Wu, D.-Y., Panneerselvam, R., and Tian, Z.-Q. (2016). Nanostructure-based plasmon-enhanced Raman spectroscopy for surface analysis of materials. *Nat Rev Mater* 1 (6): 16021.

35 Nie, S. and Emory, S.R. (1997). Probing single molecules and single nanoparticles by surface-enhanced Raman scattering. *Science (1979)* 275 (5303): 1102–1106.

36 Xu, H., Bjerneld, E.J., Käll, M., and Börjesson, L. (1999). Spectroscopy of single hemoglobin molecules by surface enhanced Raman scattering. *Phys Rev Lett* 83 (21): 4357–4360.

37 Han, X.X., Huang, G.G., Zhao, B., and Ozaki, Y. (2009). Label-free highly sensitive detection of proteins in aqueous solutions using surface-enhanced Raman scattering. *Anal Chem* 81 (9): 3329–3333.

38 Shin, H., Oh, S., Kang, D., and Choi, Y. (2020). Protein quantification and imaging by surface-enhanced Raman spectroscopy and similarity analysis. *Adv Sci* 7 (11): 1903638.

39 Ma, H., Tang, X., Liu, Y., Han, X.X., He, C., Lu, H., and Zhao, B. (2021). Surface-enhanced Raman scattering for direct protein function investigation: Controlled immobilization and orientation. *Anal Chem* 91 (14): 8767–8771.

40 Matousek, P., Towrie, M., and Parker, A.W. (2002). Fluorescence background suppression in Raman spectroscopy using combined Kerr gated and shifted excitation Raman difference techniques. *J Raman Spectrosc* 33 (4): 238–242.

41 Ngola, S.M., Zhang, J., Mitchell, B.L., and Sundararajan, N. (2008). Strategy for improved analysis of peptides by surface-enhanced Raman spectroscopy (SERS) involving positively charged nanoparticles. *J Raman Spectrosc* 39 (5): 611–617.

42 Anderson, M.S. (2000). Locally enhanced Raman spectroscopy with an atomic force microscope. *Appl Phys Lett* 76 (21): 3130–3132.

43 Hayazawa, N., Inouye, Y., Sekkat, Z., and Kawata, S. (2000). Metallized tip amplification of near-field Raman scattering. *Opt Commun* 183 (1–4): 333–336.

44 Schmid, T., Opilik, L., Blum, C., and Zenobi, R. (2013). Nanoscale chemical imaging using tip-enhanced Raman spectroscopy: A critical review. *Angew Chem Int Ed* 52 (23): 5940–5954.

45 Wessel, J. (1985). Surface-enhanced optical microscopy. *J Opt Soc Am B* 2 (9): 1538.

46 Yeo, B.S., Stadler, J., Schmid, T., Zenobi, R., and Zhang, W. (2009). Tip-enhanced Raman Spectroscopy – Its status, challenges and future directions. *Chem Phys Lett* 472 (1–3): 1–13.

47 Huang, B., Bates, M., and Zhuang, X. (2009). Super-resolution fluorescence microscopy. *Annu Rev Biochem* 78 (1): 993–1016.

48 Deckert-Gaudig, T., Taguchi, A., Kawata, S., and Deckert, V. (2017). Tip-enhanced Raman spectroscopy – From early developments to recent advances. *Chem Soc Rev* 46 (13): 4077–4110.

49 Zhu, F., Isaacs, N.W., Hecht, L., and Barron, L.D. (2005). Raman optical activity: A tool for protein structure analysis. *Structure* 13 (10): 1409–1419.

50 Richards, D., Milner, R.G., Huang, F., and Festy, F. (2003). Tip-enhanced Raman microscopy: Practicalities and limitations. *J Raman Spectrosc* 34 (9): 663–667.

51 Hayazawa, N., Watanabe, H., Saito, Y., and Kawata, S. (2006). Towards atomic site-selective sensitivity in tip-enhanced Raman spectroscopy. *J Chem Phys* 125 (24): 244706.

52 Park, K.D., Kim, Y.H., Park, J.H., Park, J.S., Lee, H.S., Yim, S.Y., Lee, Y.H., and Jeong, M.S. (2012). Ultraviolet tip-enhanced nanoscale Raman imaging. *J Raman Spectrosc* 43 (12): 1931–1934.

53 Kumar, N., Rae, A., and Roy, D. (2014). Accurate measurement of enhancement factor in tip-enhanced Raman spectroscopy through elimination of far-field artefacts. *Appl Phys Lett* 104 (12): 123106.

54 Lerman, G.M. and Levy, U. (2008). Effect of radial polarization and apodization on spot size under tight focusing conditions. *Opt Express* 16 (7): 4567.

55 Zhang, R., Zhang, Y., Dong, Z.C., Jiang, S., Zhang, C., Chen, L.G., Zhang, L., Liao, Y., Aizpurua, J., Luo, Y., Yang, J.L., and Hou, J.G. (2013). Chemical mapping of a single molecule by plasmon-enhanced Raman scattering. *Nature* 498 (7452): 82–86.

56 Stadler, J., Schmid, T., and Zenobi, R. (2010). Nanoscale chemical imaging using top-illumination tip-enhanced Raman spectroscopy. *Nano Lett* 10 (11): 4514–4520.

57 Wang, P., Zhang, D., Li, L., Li, Z., Zhang, L., and Fang, Y. (2012). Reversible defect in graphene investigated by tip-enhanced Raman spectroscopy. *Plasmonics* 7 (3): 555–561.

58 Zhang, D., Heinemeyer, U., Stanciu, C., Sackrow, M., Braun, K., Hennemann, L.E., Wang, X., Scholz, R., Schreiber, F., and Meixner, A.J. (2010). Nanoscale spectroscopic imaging of organic semiconductor films by plasmon-polariton coupling. *Phys Rev Lett* 104 (5): 056601.

59 Krasnoslobodtsev, A.V., Deckert-Gaudig, T., Zhang, Y., Deckert, V., and Lyubchenko, Y.L. (2016). Polymorphism of amyloid fibrils formed by a peptide from the yeast prion protein Sup35: AFM and Tip-Enhanced Raman Scattering studies. *Ultramicroscopy* 165: 26–33.

60 Cowcher, D.P., Deckert-Gaudig, T., Brewster, V.L., Ashton, L., Deckert, V., and Goodacre, R. (2016). Detection of protein glycosylation using tip-enhanced Raman scattering. *Anal Chem* 88 (4): 2105–2112.

61 Wang, H., Carrier, S.L., Park, S., and Schultz, Z.D. (2015). Selective TERS detection and imaging through controlled plasmonics. *Faraday Discuss* 178 (0): 221–235.

62 Jakob, T., Österlund, L., and Edvinsson, T. (2021). Polarized and non-polarized Raman spectroscopy of ZnO crystals: Method for determination of crystal growth and crystal plane orientation for nanomaterials. *J Raman Spectrosc* 52: 1395–1405.

63 Sereda, V. and Lednev, I.K. (2014). Polarized Raman spectroscopy of aligned insulin fibrils. *J Raman Spectroscopy* 45(8): 665–671.

64 Carrey, P.R. (2006). Raman crystallography and other biochemical applications of Raman microscopy. *Annu Rev Phys Chem* 57: 527–554. doi: 10.1146/annurev.physchem.57.032905.104521.

65 Rodríguez-Pérez, J.C., Hamley, I.W., and Squires, A.M. (2013). Determination of orientations of aromatic groups in self-assembled peptide fibrils by polarised Raman spectroscopy. *Phys Chem Chem Phys* 15: 13940.

66 Tsuboi, M., Ueda, T., Ushizawa, K., Ezaki, Y., Overman, S.A., and Thomas, G.J. (1996b). Raman tensors for the tryptophan side chain in proteins determined by polarized Raman microspectroscopy of oriented N-acetyl-L-tryptophan crystals. *J Mol Struct* 379: 43–50.

67 Galvis, L., Dunlop, J.W.C., Duda, G., Fratzl, P., and Masic, A. (2013). Polarized Raman anisotropic response of collagen in tendon: Towards 3D orientation mapping of collagen in tissues. *Plos One* 8: e63518–e63527.

68 Carrey, P.R. (2011). Kinetic crystallography by Raman microscopy.

69 Chen, Y., Basu, R., Gleghorn, M.L., Murakami, K.S., and Carey, P.R. (2011 August 17). Time resolved events on the reaction pathway of transcript initiation by a single subunit rna polymerase: Raman crystallographic evidence. *J Am Chem Soc* 133 (32): 12544–12555. doi: 10.1021/ja201557w.

70 Carey, P.R. and Dong, J. (2004). Following ligand binding and ligand reactions in proteins via Raman crystallography. *Biochemistry* 43: 8885–8893.

71 McGeehan, J.E., Bourgeois, D., Royant, A., and Carpentier, P. (2011). Raman-assisted crystallography of biomolecules at the synchrotron: Instrumentation, methods and applications. *Biochim Biophys Acta* 1814: 750–759.

72 Altose, M.D., Zheng, Y., Dong, J., Palfey, B.A., and Carey, P.R. (2001 March 13). Comparing protein–ligand interactions in solution and single crystals by Raman spectroscopy. *PNAS* 98 (6): 3006–3011.

73 Pelton, J.T. and McLean, L.R. (2000). Spectroscopic methods for analysis of protein Secondary structure. *Anal Biochem* 277: 167–176.

74 Noda, I. (2017). Vibrational two-dimensional correlation spectroscopy (2DCOS) study of proteins. *Spectrochim Acta A Mol Biomol Spectrosc* 187 (5): 119–129.

75 Noda, I. (1986). Two-dimensional infrared (2D IR) spectroscopy of synthetic and biopolymers. *Bull Am Phys Soc* 31: 520.

76 Ashton, L., Barron, L.D., Czarnik-Matusewicz, B., Hecht, L., Hyde, J., and Blanch, E.W. (2006). Two dimensional correlation analysis of Raman optical activity data on the α-helix-to-β-sheet transition in poly(L-lysine). *Mol Phys* 104 (9): 1429–1445.

77 Noda, I. and Ozaki, Y. (2004). *Two-dimensional Correlation Spectroscopy: Applications in Vibrational and Optical Spectroscopy*. Chichester, UK: John Wiley& Sons.

78 Shashilov, V.A. and Lednav, I.K. (2009). Two-dimensional correlation Raman spectroscopy for characterizing protein structure and dynamics. *J Raman Spectrosc* 40: 1749–1758.

79 Schultz, C.P., Bârzu, O., and Mantsch, H.H. (2000). Two-dimensional infrared correlation analysis of protein unfolding: Use of spectral simulations to validate structural changes during thermal denaturation of bacterial, CMP binases. *Appl Spectrosc* 54 (7): 931–938.

80 Czarnik-Matusewicz, B., Murayama, K., Wu, Y., and Ozaki, Y. (2000). Two-dimensional attenuated total reflection/infrared correlation spectroscopy of adsorption-induced and concentration-dependent spectral variations of β-lactoglobulin in aqueous solutions. *J Phys Chem B* 104 (32): 7803–7811.

81 Dluhy, R., Shanmukh, S., and Morita, S.I. (2006). The application of two-dimensional correlation spectroscopy to surface and interfacial analysis. *Surf Interface Anal* 38: 1481–1496.

82 Cerdà-Costa, N., De la Arada, I., Avilés, F.X., Arrond, J.L.R., and Villegas, S. (2009). Influence of aggregation propensity and stability on amyloid fibril formation as studied by Fourier transform infrared spectroscopy and two-dimensional COS analysis. *Biochemistry* 48 (44): 10582–10590.

83 Zhang, M., Dang, Y.Q., Liu, T.Y., Li, H.W., Wu, Y., Li, Q., Wang, K., and Zou, B. (2013). Pressure-induced fluorescence enhancement of the BSA-protected gold nanoclusters and the corresponding conformational changes of protein. *J Phys Chem C* 117 (1): 639–647.

84 Wu, Y., Murayama, K., and Ozaki, Y. (2001). Two-dimensional infrared spectroscopy and principle component analysis studies of the secondary structure and kinetics of hydrogen-deuterium exchange of human serum albumin. *J Phys Chem B* 105 (26): 6251–6259.

85 Zhou, P., Xie, X., Knight, D.P., Zong, X., Deng, F., and Yao, W. (2004). Effects of pH and calcium ions on the conformational transitions in silk fibroin using 2D Raman correlation spectroscopy and 13C solid-state NMR. *Biochemistry* 43 (35): 11302–11311.

86 Ashton, L., Barron, L.D., Czarnik-Matusewica, B., Hecht, L., Hyde, J., and Blanch, E.W. (2005). Two-dimensional correlation analysis of Raman optical activity data on the α-helix-to-β-sheet transition in poly(L-lysine). *Mol Phys* 104 (9): 1429–1445.

87 Ruan, Q., Zhou, P., Hu, B., and Ji, D. (2008). An investigation into the effect of potassium ions on the folding of silk fibroin studied by generalized two-dimensional NMR-NMR correlation and Raman spectroscopy. *FEBS J* 275 (2): 219–232.

88 Rodriguez-Casado, A., Molina, M., and Carmona, P. (2007). Core protein—Nucleic acid interactions in hepatitis C virus as revealed by Raman and circular dichroism spectroscopy. *Appl Spectrosc* 61 (11): 1219–1224.

89 Paquin, R. and Colomban, P. (2006). Nanomechanics of single keratin fibres: A Raman study of the α-helix →β-sheet transition and the effect of water. *Raman Spectrosc* 38 (5): 504–514.

90 Mello, C., Ciuffi, K.J., Nassar, E., Ribeiroa, D., Poppi, R., and Braz, J. (2006). Two-dimensional low resolution Raman spectroscopy applied to fast discrimination of clinically relevant microorganisms: A whole–organism fingerprinting approach. *J Braz Chem Soc* 17 (1): 73–78.

91 Ashton, L. and Blanch, E.W. (2010). pH-induced conformational transitions in α-lactalbumin investigated with two-dimensional Raman correlation variance plots and moving windows. *J Mol Struct* 974 (1–3): 132–138.

92 Barth, A. and Zscherp, C. (2002). What vibrations tell us about proteins. *Q Rev Biophys* 35: 369–430.

93 Tuma, R. (2005). Raman spectroscopy of proteins: From peptides to large assemblies. *J Raman Spectrosc* 36: 307–319.

94 Mangialardo, S., Piccirilli, F., Perucchi, A., Dore, P., and Postorino, P. (2012). Raman analysis of insulin denaturation induced by high-pressure and thermal treatments. *J Raman Spectrosc* 43: 692–700.

95 Hernández, B., Pflüger, F., Derbel, N., De Coninck, J., and Ghomi, M. (2010). Vibrational analysis of amino acids and short peptides in hydrated media. VI. Amino acids with positively charged side chains: L-lysine and L-arginine. *J Phys Chem B* 114 (2): 1077–1088.

96 Hernández, B., Pflüger, F., Adenier, A., Kruglik, S.G., and Ghomi, M. (2010). Vibrational analysis of amino acids and short peptides in hydrated media. VIII. Amino acids with aromatic side chains: L-phenylalanine, L-tyrosine, and L-tryptophan. *J Phys Chem B* 114 (46): 15319–15330.

97 Zhu, G., Zhu, X., Fan, Q., and Wan, X. (2011). Raman spectra of amino acids and their aqueous solutions. *Spectrochim Acta A Mol Biomol Spectrosc* 78 (3): 1187–1195.

98 Chuang, C.H. and Chen, Y.T. (2009). Raman scattering of L-tryptophan enhanced by surface plasmon of silver nanoparticles: Vibrational assignment and structural determination. *J Raman Spectrosc* 40 (2): 150–156.

99 Lagant, P., Vergoten, G., Fleury, G., and Loucheux-Lefebvre, M.H. (1984). Raman spectroscopy and normal vibrations of peptides Characteristic normal modes of a type-II β turn. *Eur J Biochem* 139 (1): 137–148.

100 Podstawka, E., Ozaki, Y., and Proniewicz, L.M. (2004). Part I: Surface-enhanced Raman spectroscopy investigation of amino acids and their homodipeptides adsorbed on colloidal Silver. *App Spectrosc* 58 (5): 118A–144A and 491–638.

101 Xiaojuan, Y., Gu Huaimin, G., and Jiwei, W. (2010). Surface-enhanced Raman spectrum of Gly-Gly adsorbed on the silver colloidal surface. *J Mol Struct* 977 (1–3): 56–61.

102 Sjöberg, B., Foley, S., Cardey, B., and Enescu, M. (2014). An experimental and theoretical study of the amino acid side chain Raman bands in proteins. *Spectrochim Acta A Mol Biomol Spectrosc* 128: 300–311.

103 Ferrer, E.G., Gomez, A.V., Anon, M.C., and Puppo, M.C. (2011). Structural changes in gluten protein structure after addition of emulsifier. A Raman spectroscopy study. *Spectrochim Acta A* 79: 278–28.

104 Robens, J. and Cardwell, K. (2003). The costs of mycotoxin management to the USA: Management of Aflatoxins in the United States. *Toxin Rev* 22: 139–152.

4

Fundamental Principles of Impedance Spectroscopy and its Biological Applications

Yusuke Tsutsui

Department of Molecular Engineering, Graduate School of Engineering, Kyoto University, Kyotodaigaku-katsura, Nishikyo-ku, Kyoto, Japan

4.1 Introduction

The term "spectroscopy" originates from the word "spectrum," meaning spatially dispersed light through a dispersing media such as a prism. From the 18th century, beginning with *Opticks* by Newton, understanding of the interaction between light and matter—emission and absorption of specific wavelengths—has been broadly studied in the spectral range spanning from X-ray to radio frequency. As shown in Figure 4.1, since spectroscopy itself is quite a broad field, as can be easily imagined, related physical phenomena, measurement methods, and the terms are totally different in each frequency range. Impedance spectroscopy is located at relatively low frequency in this sense, spanning from pseudo-DC (such as microhertz) to the mega- to-gigahertz region, which is accessible by electrical technology.

4.1.1 Basic Concept of Impedance Spectroscopy

Impedance spectroscopy is basically one of the electrical measurement techniques. The term "spectroscopy" is borrowed from the optical measurement since impedance spectroscopy similarly separates and analyzes frequency components. It is mainly used to analyze or understand the behavior and internal equivalent circuit of an unknown system by assuming a combination of passive electrical components. The components will be chosen properly to describe the response of the systems we are interested in, especially in the frequency domain (it is the key objective to check the frequency dependence of the response, as we will see soon). In other words, the black box will be replaced with the combination of electrical components under proper assumptions (Figure 4.2). The response of the system has been often analyzed by fitting with a model electrical circuit that is considered to be physically reasonable. From this assignment, it is possible to estimate the

Analytical Techniques for the Elucidation of Protein Function, First Edition. Edited by Isao Suetake, Rohit K. Sharma, and Hironobu Hojo.

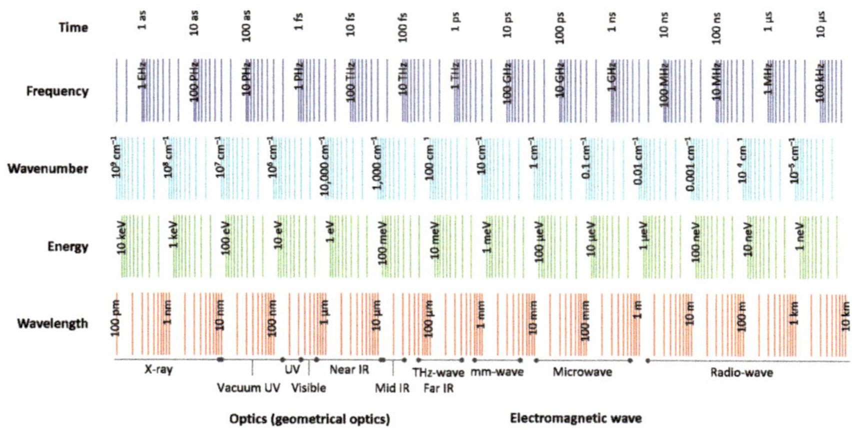

Figure 4.1 Overview of light in frequency domain with several units.

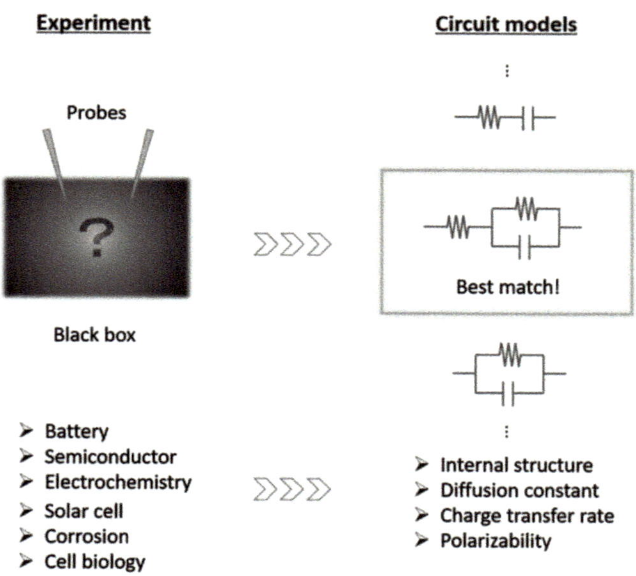

Figure 4.2 Schematic concept of impedance spectroscopy.

contributions from each physical phenomenon to understand the system more specifically and in detail, for example, to extract bulk reorganization, charge transport in the bulk, interfacial chemical reaction or charge transfer, polarizability of the material, diffusion constant of solutes, or geometrical change.

Since the stimuli should have finite frequency, we should consider the alternative current. When the sinusoidal external stimuli (such as voltage or force) is applied to the system, it is highly expected to get a response also in sinusoidal fashion (e.g. in the form of current or displacement). As we know well, there are a lot of situations that the system has totally nonlinear response, but here we limit ourselves only to the linear cases. Under this situation, these behaviors can be mathematically written in the form of

$$V = V_0 \sin\left(\omega_0 t\right), \quad I = I_0 \sin\left(\omega_0 t + \theta\right), \tag{4.1}$$

where the voltage with an amplitude (peak-to-peak: pp) of V_0 and a frequency of ω_0 has been applied, and the current flows with an amplitude of I_0 and a frequency of ω_0. The additional term θ represents the phase delay in the current, which is sometimes faster or slower than the applied voltage. The impedance spectroscopy basically measures two values, the ratio V_0/I_0 and θ, as functions of frequency.

Impedance spectroscopy is worth to learn because it offers a lot of benefits:

1) Information on the components of the system, extraction of internal physical phenomena

As we discussed above, it is highly effective with an unknown system in order to understand its overall behavior and response. In addition, it breaks down the system in smaller parts to obtain a better understanding or to extract each internal property of the system.

2) Wide applicability

Since impedance spectroscopy is basically an electrical technique, it has versatile applicability and compatibility. For example, it can be applied to characteristics of the electronic parts, photovoltaic cell, batteries, fuel cell, biosensors, corrosion, semiconductors, paintings, and food analysis.

3) Sometimes much higher sensitivity compared to DC-based techniques

From the broader viewpoint that impedance spectroscopy is an AC-based technique, it may offer higher sensitivity. As noted above, we made an assumption that the frequency of the system is the same as that of the applied stimuli. This assumption is highly effective to eliminate the noise with different frequency components, and technically it is called as lock-in detection method. In addition, it is a clear advantage that more information can be extracted compared to the DC (non-time resolved) technique.

Since it can easily be understood from the basic idea, we will start reviewing the passive electrical components and circuits. No special knowledge is required for college students to understand impedance spectroscopy. Impedance is the mathematical expansion of resistance. When you describe a resistance, the well-known Ohm's law can be used:

$$R = \frac{V}{I}, \tag{4.2}$$

where R, V, I represent a resistance, voltage, and current, respectively. Here, each variable is expected to be a real number. To understand impedance spectroscopy, we need to understand complex numbers since Ohm's law can be extended to complex number as follows by simply replacing the resistance R with the impedance Z as

$$Z = \frac{V}{I}, \tag{4.3}$$

which is much more convenient to describe the electrical response of the system of our interest. Going back to the original formula shown above,

$$V = V_0 \sin\left(\omega_0 t\right), \quad I = I_0 \sin\left(\omega_0 t + \theta\right), \tag{4.4}$$

we note that two components are needed and are actually enough to describe the response of the system. Here, what we mean by "enough" is that if the two parameters are given, the response (current) waveform can be written down without any ambiguity, as shown in Figure 4.3. What if we draw the relationship between voltage and current by a different method? For example, taking a two-dimensional plane and placing the vector with amplitude V_0 in the x-axis, the current vector can be placed with an angle of θ and a length of I_0 to the voltage vector without any ambiguity. This kind of representation is great because it can be done by a single complex number (which has two degrees of freedom, i.e. real and imaginary parts):

$$c = a + bj, \tag{4.5}$$

where c is a complex number, and a and b are real numbers representing the real and imaginary part of c, respectively. Normally the letter i is used in mathematics to represent the imaginary unit, but to avoid confusion with the current, we follow the convention to use the letter j as an alternative. Taking the x- and y-axis as real and imaginary part, respectively, the complex number c can be represented by a single dot on the plane with coordinates $(x, y) = (a, b)$. The coordinates are related to the amplitude I_0 and the angle θ with the positive x-axis.

$$a = I_0 \cos\theta, \quad b = I_0 \sin\theta, \tag{4.6}$$

where we can see a one-to-one correspondence with the representation based on the amplitude I_0 and angle θ. Hence, we can relate a complex voltage V to a complex current I by a single complex number impedance Z as shown in (4.3). We should also recall Euler's equation:

$$e^{\pm j\theta} = \cos\theta \pm j \sin\theta. \tag{4.7}$$

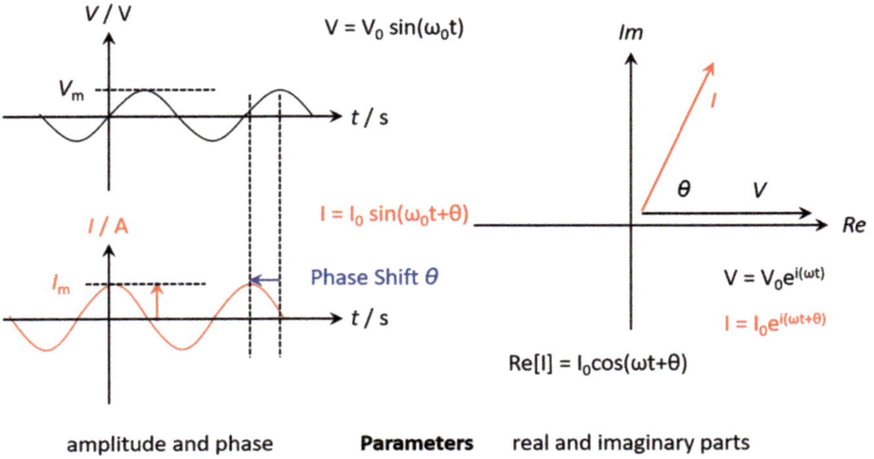

Figure 4.3 Parameters to describe the relationship between voltage and current.

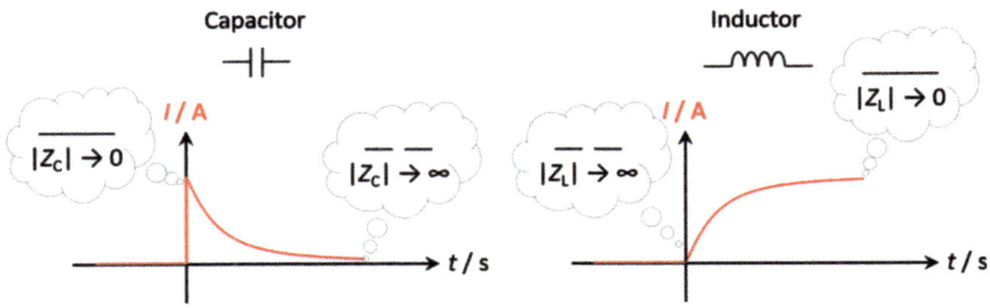

Figure 4.4 Schematic illustration of current responses of capacitors and inductors to DC step bias. Note that a finite serial resistance has been considered.

4.1.2 Description of Impedance for Capacitors and Inductors

We already know that capacitors and inductors are frequency-dependent components (or in other words, time-dependent components). For example, the current response changes by time upon application of a DC bias voltage (Figure 4.4). In the case of a capacitor, a large current flows initially to charge it, and then it decreases to reach ideally zero after sufficient time. On the other hand, an inductor has initially a large resistance and behaves later as just a conducting wire. These behaviors should be incorporated in the complex version of Ohm's law.

The relation between current and voltage is described by

$$I = C\frac{dV}{dt}, \quad V = L\frac{dI}{dt}, \tag{4.8}$$

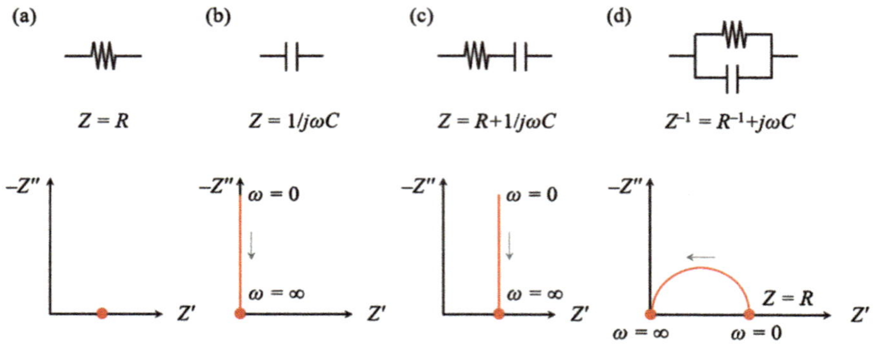

Figure 4.5 Nyquist plots for simple RC model circuits.

for capacitors and inductors, respectively, where C and L represent capacitance and inductance. Although we will not dig in detail, the differential operation can be replaced by the multiplication of imaginary the unit and the frequency, $j\omega$. So the following relationship can be obtained:

$$V = \frac{1}{j\omega C} I, \quad V = j\omega L I. \tag{4.9}$$

Now we obtain the description of impedance for the passive components, resistors (Z_R), capacitors (Z_C), and inductors (Z_L), as

$$Z_R = R, \quad Z_C = -j\frac{1}{\omega C}, \quad Z_L = j\omega L, \tag{4.10}$$

where the impedance of resistances is apparently the same as the resistance R.

4.1.3 Nyquist Plot

There is a useful and convenient way to graphically express the impedance: a Nyquist plot (also called Cole-Cole plot). It is exactly the impedance plotted in the complex plane. Since the impedance changes by scanning the frequency from zero to infinity, an "impedance point" written on the complex plane moves around accordingly to draw a locus.

This type of figure has been used frequently because we can quickly determine what kind of components are included in the system, and it provides a comprehensive view in a wide range of frequencies with only a single graph. Examples of the Nyquist plot are shown in Figure 4.5 for several basic model circuits, where Z' and Z'' represent Re[Z] (real part of Z) and Im[Z] (imaginary part of Z), respectively. By convention, the negative direction is chosen in the figure as the direction of the y-axis. We can easily recognize their difference in their Nyquist plots. In circuit (a), it contains only one bullet on $Z' = R$ since we consider a single resistance R (which is apparently frequency independent). A single capacitor only contributes to the imaginary part, and it changes from negative infinity to 0 with the frequency scanned from 0 to infinity, as shown in (b). Here the arrow indicates the increase in the frequency. In this manner, the frequency doesn't appear explicitly in the Nyquist plot. Figure (c) can be understood similarly: The addition of the single resistor R shifts the line in (b) by R. Figure 4.5(d) represents an RC parallel circuit, apparently different from the others. As practice, let's consider more details on this circuit to get better insights on Nyquist plots. Although these are the model circuits, the basic understanding of them nourishes our intuitions for future analysis of real-world cases.

In the RC parallel circuit the total impedance Z can be represented as:

$$\frac{1}{Z} = \frac{1}{Z_R} + \frac{1}{Z_C} = R^{-1} + j\omega C. \tag{4.11}$$

Inversion is taken to obtain

$$Z = \frac{1}{R^{-1} + j\omega C} = \frac{R^{-1} - j\omega C}{R^{-2} + (\omega C)^2} = \frac{R - j\omega C R^2}{1 + (\omega C R)^2} = \frac{R}{1 + (\omega C R)^2} - j\frac{\omega C R^2}{1 + (\omega C R)^2}. \tag{4.12}$$

Since the total impedance is now expressed in the form of a complex number, it can be drawn on the complex plane. (In the current framework, any complex circuit can be expressed with complex numbers, i.e. a pair of real numbers because of the algebraic closure.) Considering that a Nyquist plot describes the relationship between the real and imaginary parts of the impedance and that frequency dependence is incorporated implicitly, we should relate them by removing ω from Equation (4.12). Namely, the following equation should be reformulated:

$$\begin{cases} Z' = \dfrac{R}{1 + (\omega C R)^2} \\ Z' = -\dfrac{\omega C R^2}{1 + (\omega C R)^2} \end{cases} \Rightarrow \begin{cases} (\omega C R)^2 = \dfrac{R}{Z'} - 1 \\ Z'^2 = \dfrac{(\omega C R)^2 R^2}{\left(1 + (\omega C R)^2\right)^2} \end{cases}. \tag{4.13}$$

Eliminating the frequency ω gives

$$Z'^2 = \frac{\left(\dfrac{R}{Z'} - 1\right) R^2}{\left(\dfrac{R}{Z'}\right)^2} = \left(\frac{R}{Z'} - 1\right) R^2 \left(\frac{Z'}{R}\right)^2 = RZ' - Z'^2. \tag{4.14}$$

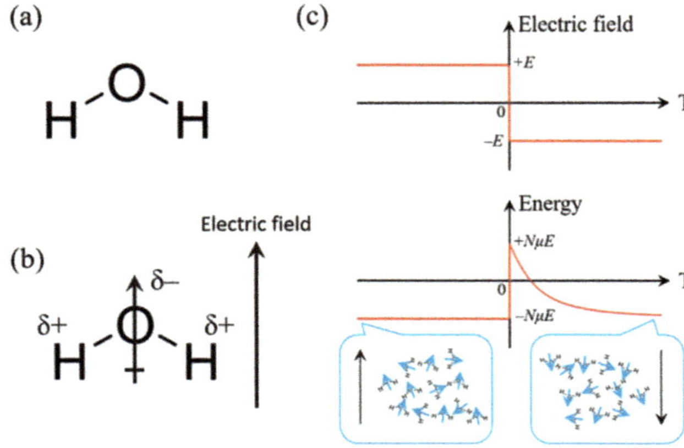

Figure 4.6 Water molecules trying to minimize the energy of the system. (a) Molecular structure of water, (b) water molecule with its dipole moment. The variable δ expresses the partial distribution of the electrons and satisfies $0 \leqslant \delta \leqslant 1$. (c) The schematic illustration of the electric field response of the system containing a lot of water molecules. E, N, and μ represent a maximum electric field applied to the system, the amplitude of the dipole moment for one water molecule, and the magnitude of the dipole moment, respectively.

By further reformulation, the following equation is obtained:

$$\left(Z' - \frac{R}{2}\right)^2 + \left(Z'\right)^2 = \left(\frac{R}{2}\right)^2. \tag{4.15}$$

This clearly represents the equation of a circle on the complex plane with a radius of $R/2$ and center $(Z', Z'') = (R/2, 0)$. Considering that the Z'' is limited to non-positive numbers [from Eq. (4.13)], it is a semicircle, as shown in Figure 4.5(d). From Equation (4.12), $\omega \to 0$ and ∞ represent $Z = R$ and 0, respectively. It is also reasonable from the standpoint of the electrical circuit since a capacitance can be recognized as an insulator at low frequency limit ($Z = R$) and a conducting wire at high frequency limit ($Z = 0$). We didn't mention inductors here, but it can also be considered in a similar way.

As we have seen in this section, circuit information could be extracted from a Nyquist plot even if we face a black box (unfortunately, the whole information not always can be obtained. It depends on the situation. For example, if two parameters are too close, it will be difficult to separate them). In the next chapter, we will consider the application of this method in real responses of the media.

4.1.4 Debye Model

Apart from the simple circuit model, let's consider the response of the material. Water is known to have a static dielectric constant of $\varepsilon_s \sim 80$ and in a molecular sense it has a dipole moment of ~1.9 Debye originating from the partial negative

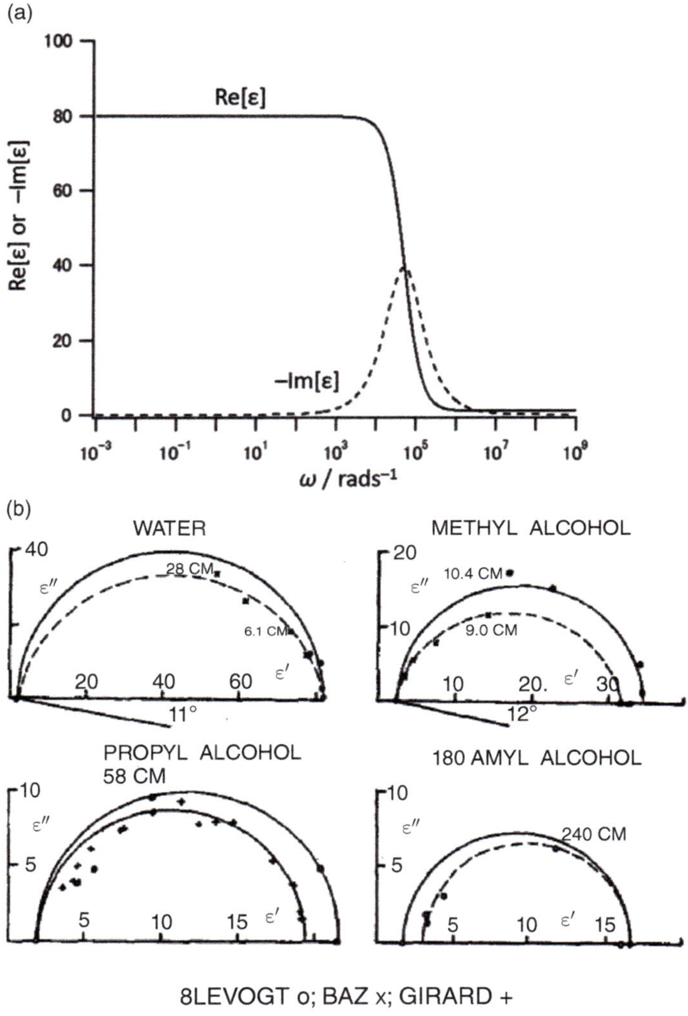

Figure 4.7 (a) Simulated real and imaginary parts of Debye dispersion with parameters of 1 MHz to 1 GHz, $\varepsilon_s = 80$, $\varepsilon_\infty = 1$, $\tau_0 = 2 \times 10^{-5}$ s. (b) Cole-Cole plot of various dielectrics (Cole et al. 1941 / With permission of AIP Publishing).

(positive) charge on the oxygen (hydrogen) atom (Figure 4.6(a) and (b)). Due to this dipole moment, the system acquires an energy difference under an external electric field E upon molecular orientation. Water molecules orient along the direction of an external electric field to minimize the energy of the system (for sure, we have a randomness or distribution in the angle due to the thermal fluctuation). As shown in Figure 4.6(c), if we consider the situation that the external electric field has been reversed suddenly from one direction to the opposite direction, the system becomes unstable and water molecules try to reorient themselves. Since the atoms should move, it requires some amount of time to completely relax to the most stable state, which is called relaxation time (τ_0). This situation can be considered in the frequency domain where the water molecule has enough time to respond to the external alternating electric field with low frequency, but it's not possible under that of high frequency. This means that the dielectric constant is a frequency-dependent parameter and so it's better to represent it as $\varepsilon(\omega)$.

This relaxation behavior was formulated by P. Debye as

$$\varepsilon(\omega)=\varepsilon_\infty+\frac{\varepsilon_s-\varepsilon_\infty}{1+j\omega\tau_0}, \tag{4.16}$$

where ε_∞ is the dielectric constant at the high-frequency limit. It can be written in a separated form as

$$\varepsilon(\omega)=\varepsilon_\infty+\frac{\varepsilon_s-\varepsilon_\infty}{1+(\omega\tau_0)^2}-j\frac{(\varepsilon_s-\varepsilon_\infty)\omega\tau_0}{1+(\omega\tau_0)^2}. \tag{4.17}$$

The equation is visualized in Figure 4.7. The real part shows monotonic decrease from ε_s to ε_∞ upon increase in the frequency. On the other hand, the imaginary part shows peak-shape, which converges to 0 at both the high- and low-frequency limits. Note that this equation is close to the one we discussed in the parallel RC circuit model [refer to Eq. (4.12)]. If we replace the variable *CR* with τ_0, *R* with $\varepsilon_s - \varepsilon_\infty$, the difference is only the first term ε_∞ in the real part. Since this minor difference only shifts the semicircle in the complex plane toward the *x*-axis, Equation (4.17) still represents a semicircle in the complex plane. The actual behavior of various dielectrics is shown in Figure 4.7.

Beyond the simple Debye model [1], several descriptions have been proposed such as the Cole-Cole eq. [2], Cole-Davidson eq. [3], and Havriliak–Negami eq. [4]. These models improve asymmetricity and distortion observed in real dielectrics, which cannot be described by the simple Debye model.

Especially, we focus on empirically derived Havriliak–Negami relaxation because it explicitly contains both Cole-Cole and Cole-Davidson models and it is an extension of the Debye relaxation model. It was first used to explain the dielectric properties of polymer materials with the following formula:

$$\varepsilon(\omega)=\varepsilon_\infty+\frac{\varepsilon_s-\varepsilon_\infty}{\left(1+(j\omega\tau_0)^\alpha\right)^\beta},\quad 0\le\alpha,\beta\le 1. \tag{4.18}$$

The Cole-Cole and Cole-Davidson relations are equivalent to the Navrilak–Negami relation with $\beta = 1$ and $\alpha = 1$, respectively. While the Cole-Cole relation

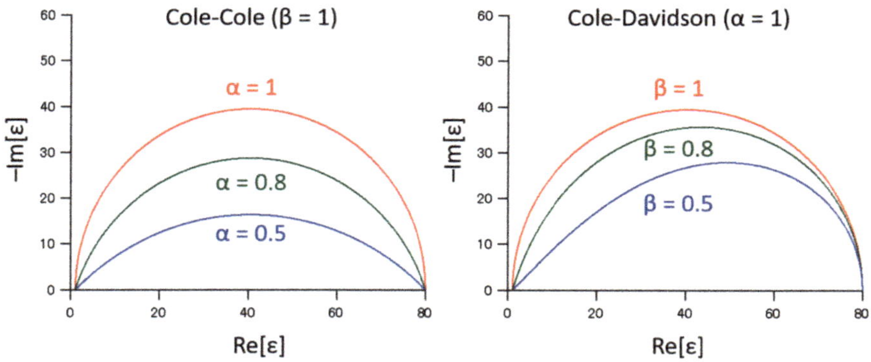

Figure 4.8 Cole-Cole and Cole-Davidson plots for different parameters with frequency range of 1 MHz to 1 GHz, $\varepsilon_s = 80$, $\varepsilon_\infty = 1$, $\tau = 2 \times 10^{-5}$ s.

can reproduce distortion along the imaginary axis and is mainly used to model liquids and biological tissues, Cole-Davidson relation generates asymmetricity on the real axis, as shown in Figure 4.8. As shown in the above, the dielectric response often draw a distorted semicircle, which could be modeled by a modified parallel RC circuit. In the next chapter we will discuss how to deal with the distorted semicircle from the equivalent circuit model.

4.1.5 Constant Phase and Warburg Element to Model Distorted and Diffusive Components

Coming back to the model analysis using electrical components, in the real analysis, a semicircle often appears with a distorted shape. The reason for this behavior has been considered to be inhomogeneity of surface, current density or distribution, and liquid composition. Normally it is treated by replacing the capacitive component with a constant phase element (CPE) with an impedance of

$$Z_{CPE} = \frac{1}{(j\omega)^p C_{CPE}}, \tag{4.19}$$

where C_{CPE} and p $(0 \le p \le 1)$ represent the CPE constant and exponent, respectively. CPE becomes a normal capacitor with $p = 1$, where the constant C_{CPE} is equivalent to the capacitance C. The above equation can be represented in a separated form as

$$Z_{CPE} = \frac{1}{\omega^p C_{CPE}}\left(\cos\left(\frac{\pi}{2}p\right) - j\sin\left(\frac{\pi}{2}p\right)\right), \tag{4.20}$$

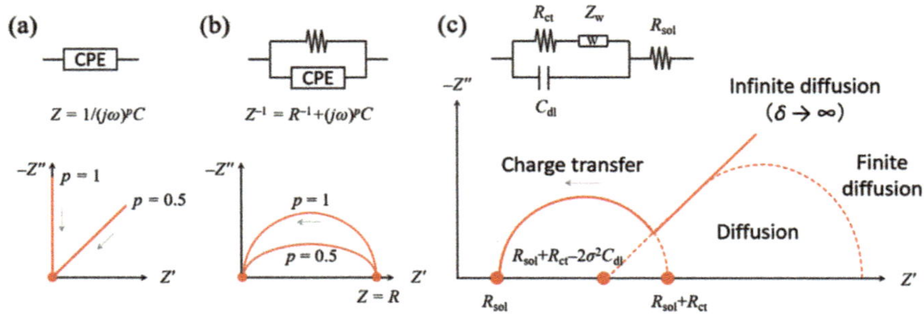

Figure 4.9 Nyquist plots for model circuits with (a, b) CPE and (c) Warburg impedance (Randles circuit).

or alternatively, we can rewrite it in the form of amplitude and phase as

$$|Z| = \frac{1}{\omega^p C_{CPE}}, \quad \theta = -\frac{\pi}{2}p, \tag{4.21}$$

which is the reason why this component is called a constant phase element (since θ is constant for the given p). As shown in Figure 4.9, the Nyquist plot of CPE shows a single straight line itself or distorted semicircle when it forms a parallel circuit with a resistance. Especially, the line has an angle of 45 degrees with the x-axis with $p = 0.5$, which sometimes appears in the low frequency region and is related to the diffusion of reactants in real media. It's called Warburg impedance [5] and is expressed with the following formula:

$$Z_W = \left(\frac{\sigma}{\sqrt{\omega}} - j\frac{\sigma}{\sqrt{\omega}}\right)\tanh\left(\delta\sqrt{\frac{j\omega}{D}}\right), \tag{4.22}$$

where δ is the thickness of the diffusion layer and D is the diffusion constant of the reactant. The parameter σ, the slope along $\omega^{-1/2}$, stands for

$$\sigma = \frac{RT}{\sqrt{2}n^2F^2A}\left(\frac{1}{\sqrt{D_O}c_O} + \frac{1}{\sqrt{D_R}c_R}\right). \tag{4.23}$$

It contains the diffusion constant and concentration of oxidant D_O and c_O (reductant D_R and c_R) with values of Faraday constant F, gas constant R, area of the electrode A, and the number of electrons involved in the reaction. The tanh term in (4.22) converges to 1 when δ is large enough (semi-infinite diffusion). Warburg impedance is often introduced with serial to R_{ct} in the parallel RC circuit, as shown in Figure 4.9 to describe the system impedance with the electrochemical reactions (Randles circuit).

4.2 Biological Applications of Impedance Spectroscopy

4.2.1 Detection of DNA Hybridization and Photodamage

Detection and identification of a specific oligonucleotide sequence is important for viral infections, gene mapping and expression, and clinical and forensic applications. It can be used conveniently to identify certain infections and diseases. Although sequencing techniques such as polymerase chain reaction (PCR) have been well developed recently (it requires well trained technicians to properly operate), DNA hybridization is one of the classical techniques that is applied to the DNA microarray and fluorescence in situ hybridization (FISH) method to detect a complemental sequence with high throughput or to achieve spatial resolution in situ. Biosensors based on DNA hybridization have been developed where the short single strand oligonucleotide probes were tethered on the electrode to make a pair binding to complemental DNA [6] (Figure 4.10). Several electrochemical methods have been reported with acoustic waves [7, 8], surface plasmon resonance [9, 10], piezoelectric detectors [11], and capacitance detectors [12].

Labeling free measurements has been developed with impedance sensing by functionalized insulators, semiconductors, and heterostructures. When the metal electrodes are used as a substrate for DNA hybridization, a sensitivity several orders of magnitude higher, e.g. ~1 pg/ml, has been achieved in electric measurements [12]. For example, attached ssDNA on a gold surface tends to lie onto the surface or self-folds due to its flexibility. Binding to the complemental

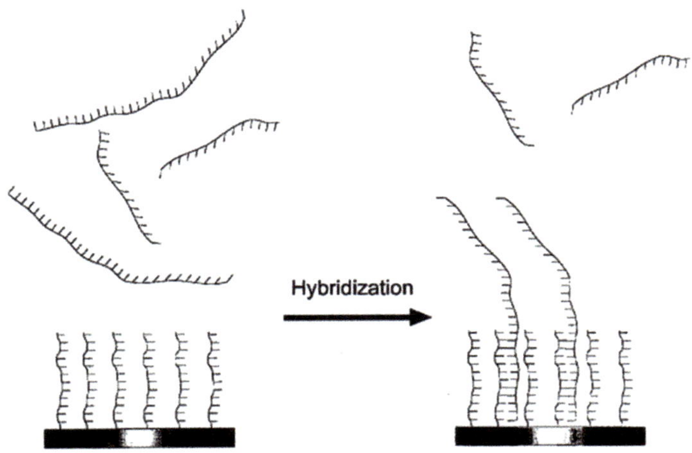

Figure 4.10 Schematic illustration of DNA hybridization technique by electrical measurement (Berggren et al. 1999 / With permission of John Wiley & Sons).

DNA forms a rigid backbone and detaches the nucleobase from the surface. A negatively charged phosphate backbone repels freely diffusing redox species such as $[Fe(CN)_6]^{3-/4-}$, which is also negatively charged, resulting in the detection of the increase in the charge transfer resistance [13]. In order to enhance the sensitivity, thiol co-decoration has been reported [14]. Since ssDNA is not densely adsorbed, post-adsorption of thiol displaces loosely attached ssDNA from the surface. The facile access of redox species toward the electrode results in the lower charge transfer resistance due to the tunneling current through the thiol molecule. Further, intercalation of an indicator molecule has been used to detect DNA hybridization. For example, actinomycin D and proflavine are known to intercalate only dsDNA, dominantly at the G–C base pair position [15], which modulate the impedance of dsDNA itself, as shown in Figure 4.11.

This kind of techniques has been extended to detect photodamage of DNA. Since a nucleobase, a component of DNA and RNA, has optical absorbance at the ultraviolet region, the excited state is generated by light exposure, resulting in photo-damage such as dimerization of pyrimidine (cyclobutane-pyrimidine dimers and 6–4 photoproducts). Recent experiments and theoretical calculations have progressively revealed an ultrafast relaxation pathway for excited state dynamics on DNA/RNA, including transient absorption spectroscopy [16–18], liquid jet photoelectron spectroscopy [19], and fluorescence up-conversion spectroscopy. From these investigations, it is widely accepted that the relaxation occurs in sub-picosecond to picosecond timescales in nucleosides in their solution states. As examples of other types of DNA damages, several patterns have been recognized such as base pair mismatches or loss, base lesions (deamination, depurination, oxidation by reactive oxygen species, alkylation), crosslinking, and strand breaking.

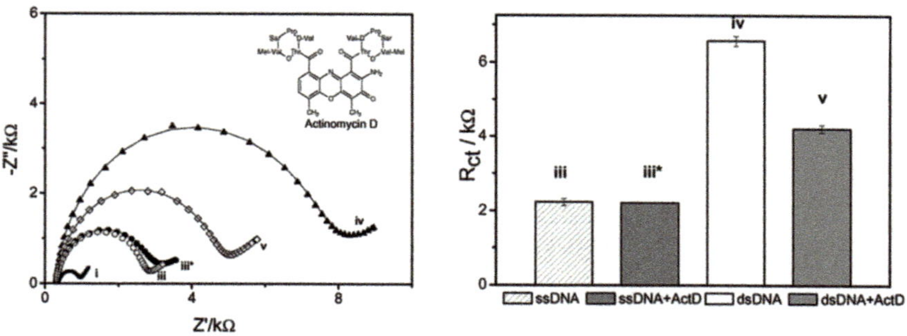

Figure 4.11 Detection of DNA-hybridization by intercalation of actinomycin D. (left) The impedance Nyquist plots of (i) bare Au electrode, ssDNA/thiol-modified Au electrode (iii) without and (iii*) with actinomycin D. dsDNA/thiol modified Au electrode (iv) without and (v) with actinomycin D. (right) Charge transfer resistance for each conditions (Gabela et al. 2009 / With permission of John Wiley & Sons).

Detection of DNA damage has been studied mostly based on electrophoretic methods, immunological methods, PCR, and chromatography. Using the intercalator technique, it was reported that DNA damaged by UV-C irradiation and reactive oxygen species showed significant decrease in R_{ct} under the presence of thioridazine. The validity of the impedance method was verified with different methods such as redox mediator method and square-wave voltammetry, indicating that the structural change of the helix was induced by the intercalation of thioridazine, which facilitates electron transfer to the electrode surface and guanine moiety to be oxidized [20]. The effect of reactive oxygen species on the dsDNA damage was also investigated by using the Fenton agent where the oxidation of guanine and adenine is accelerated in the initial treatment followed by decrease due to deeper DNA damage [21]. It was applied to investigate the damage induced by acrylamide and its metabolite [22], dopamine and metallic ions [23], human immunodeficiency virus (HIV) with concentrations from 0.5 fmol/L to 10 nmol/L [24], formation of cylclobutane pyrimidine dimers by UV-C (254 nm) and subsequent repair processes by DNA photolyase under UV-A (365 nm) photoexcitation [25]. These types of methods rather focus on the changes in the interfacial resistance but by using an interdigitated gold electrode sometimes coated by an insulator, capacitive measurement can also be done, which reflects the bulk response between the electrodes. It has been applied to the detection of neurotransmitters and toxins, acetylcholine, and antibodies.

4.2.2 Detection and Analysis of Proteins

An amyloid fibril protein is a protein that forms insoluble fibrils, mainly located outside of the organs and tissues in vivo due to its folding and aggregating nature. It shows green birefringence under the stain Congo Red. In 2016, the number of known human proteins that form β-sheet fibril structures has increased to thirty-six [26].

Amyloid-β (Aβ) and islet amyloid polypeptide (IAPP, also called amylin) are examples for these amyloid fibril proteins and are associated with Alzheimer's disease and type 2 diabetes, respectively. Type 2 diabetes can be seen in humans, cats, and dogs, related to the formation of amyloid fibrils. However, in mice negative to type 2 diabetes, such fibril structure is not observed, suggesting the correlation between formation of amyloid fibril and diabetes. In diabetes, it is known that pancreas β-cells synthesizing insulin are killed by the aggregating nature of these amyloids.

In spite of its importance a detailed understanding of the structure of the aggregation has been difficult for a long time due to its multi-scale nature [27, 28]. Especially in length scale, it spans from the atomic to micrometer resolution. A recent study has revealed the detailed structure of the aggregate state. For example, aggregation of human IAPP proceeds as the following steps. First, the protein backbone is aligned to form a β-sheet with a displacement of 4–5 Å. Second, these β-sheets stack in face-to-face fashion with a spacing of 8–10 Å, resulting in the formation of two-sheet nucleus or protofilament, where the fibril axis is perpendicular to the stacking direction. Finally, such protofilaments, typically two to eight of them, assemble into a fibril [29, 30] (Figure 4.12). It sometimes creates different polymorphs, both in the molecular and interprotofilament levels [31, 32]. Similarly, the aggregation procedure of Aβ has also been studied recently [33–35].

The aggregation has been studied by several means (Figure 4.13). For example, the introduction of a "probe" in the target protein is frequently used as a basic method where one can either chemically label the protein itself or attach the probe with an intermolecular force. The fluorescent probe embedded at the target site changes its quantum yield and/or wavelength depending on the environment of the emissive molecule. With this technique, the folding dynamics and the site-specific information such as hydrophobicity/hydrophilicity, pH, and distance can be assessed. In the background of fibril formation, thioflavin-T is the most widely used standard for this kind of measurement [36]. The free radical probe method is also a similar approach and precise distance can be obtained by one of the pulsed techniques, such as electron-electron double resonance (DEER). These techniques require structural modification, which is highly expected to distort the dynamics of the original target protein. In the case of an indirect attachment method, it would be sometimes disadvantageous that the probe can interact with different sites or stages of aggregates. Alternatively, nuclear magnetic resonance (NMR) can also provide precise local information. As for the global information, dynamic light scattering (DLS) can be used to track the time evolution of the size of the aggregate. More recently, the cryo-TEM technique has been highly developed to unveil the overall structure of the target protein under the cryogenic temperature. Circular dichroism also provides us with molecular information based on its chirality. Scientists use these techniques comprehensively maximizing their own advantages to deal with a whole picture of protein dynamics.

Compared to the structural analysis, however, the electrical property of their conformational changes is rather less understood. In the past, there were debates on the origin of the electrical conductivity of proteins. The protein aggregate is expected to be surrounded by the water using hydrophilic residues, and in fact, the loading of water molecules modulate the electrical conductivity of the protein [37, 38]. Since the formation of fibrils may modulate the system impedance due to the change in physical phenomena such as charge transport, electric polarization, and dielectric loss, multiple fibrillization steps could be captured by impedance spectroscopy.

Recently, proton conduction has been proposed as the origin of the change in impedance. It was monitored by impedance spectroscopy to obtain a better insight into the different stages of aggregation. Investigation of activation energy of conductivity in human IAPP compared to Aβ implies enhanced transport of protons

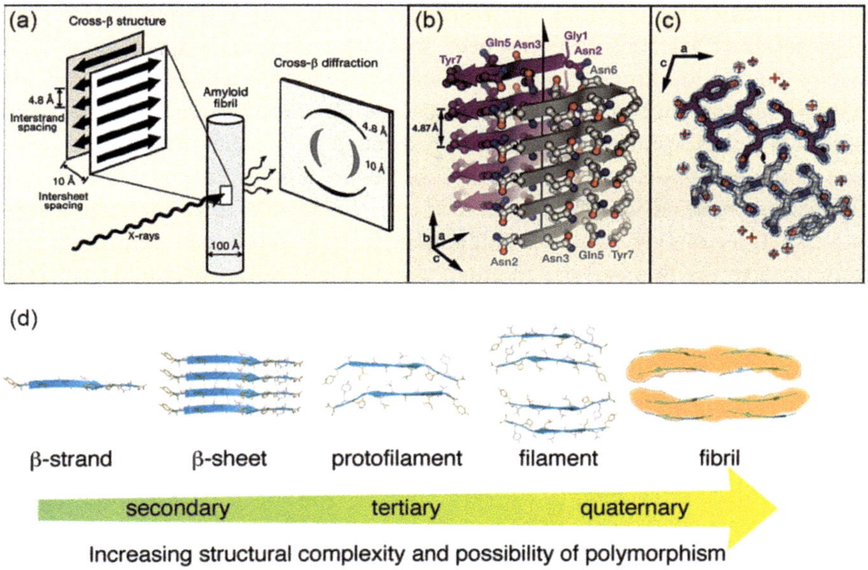

Figure 4.12 Aggregated fibril structure of the sequence segment GNNQQNY from the yeast prion Sup35 (reprinted from Eisenberg et al., *Cell*, 2012) and the eleven-residue fragment of the protein transthyretin, TTR (105–115) amyloid (adapted from Fitzpatrick et al. 2013).

	Sensitivity	Convenience	Real-time	Area	Characteristics
Circular dichroism	△	○	◎	Broad	Chirality
NMR	×	×	×	Local	Superconducting magnet
Fluorescent, radical probe	◎	△	△	Local	Introduction of probe molecule, electromagnet
Dynamic light scattering	○	△	×	Non-local	Size of the aggregates and colloids
Neutron scattering	×	×	×	Broad	Neutron facility
Impedance	○	◎	○	Non-local	Introduction of electrodes

Figure 4.13 Several experimental techniques to investigate protein dynamics.

through interstrand hydrogen bonds. For Aβ, the growth along the fibril direction dominates the aggregation process followed by the growth along the perpendicular direction. In human IAPP, however, it is shown that monomeric states are stabilized and the protofilament grows in a perpendicular direction in the earlier stage [39]. Four probe measurement of micrometer-long crystalline tyrosines revealed hole transport due to proton rocking and the energetics of proton acceptor was the critical parameter for transport rate [40].

Although the carrier transport mechanism has not reached a consensus yet, the application of impedance spectroscopy has been developed to detect quantitatively tiny amounts of target compound. For example, irreversible aggregation of human insulin has been studied by impedance spectroscopy and compared with the conventional fluorescent method [41]. The cell was modeled with a combination of a serial RC and a parallel RC circuit to describe interfacial and bulk polarization/charge transfer, respectively. The time trace of the resistive component followed a sigmoidal behavior, similar to the trend of the fluorescent method without addition of dye, which may alter the aggregation process. Interestingly, the CPE component in bulk showed more prompt rise compared to the fluorescence method, which might reflect the sudden change and saturation in charge transport behavior in the latter aggregation stage. It is known, in fact, that thioflavin-T can bind several stages of aggregation including oligomers and protofibrils [42].

In order to differentiate the contribution from each aggregation stage, a neuronal receptor cellular prion protein (PrP^C) that binds specifically to Aβ oligomers

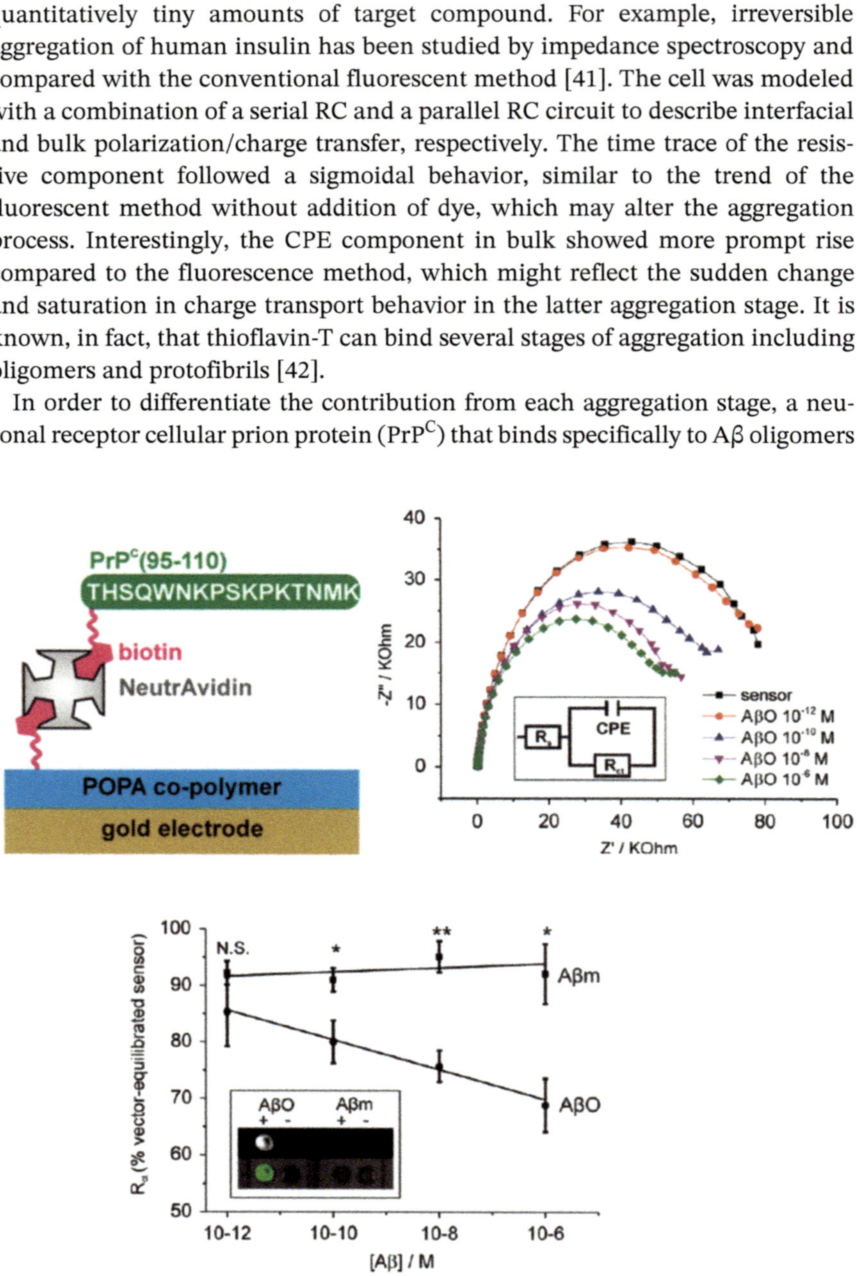

Figure 4.14 Detection of Aβ oligomers by impedance spectroscopy [43], Rushworth et al. 2014 / With permission of Elsevier.

but not to the monomers and fibrils has been utilized [43]. As shown in the Figure 4.14, the substrate with gold electrodes was coated by co-polymer, and PrP^C was tethered on the co-polymer via high affinity biotin/NeutrAvidin interaction. The impedance of the sensor has decreased after the treatment with specific concentration of Aβ oligomer, and fitting results showed that the R_{ct} has a clear negative correlation to the Aβ oligomer concentration. By using R_{ct} as an indicator, the sensor has been reported to have a detection limit of ~0.5 pM for the Aβ oligomer. In addition, it specifically detects oligomers and does not show significant response to the monomer state. Similar methods have been utilized to detect anti-transglutaminase antibodies [44], lysozyme [45], lead [46], short DNA strand HIV [24, 47], and *E. coli* [48].

4.3 Conclusion

This chapter has presented a fundamental introduction to impedance spectroscopy and its biological applications. Since impedance spectroscopy is based on electrical measurements, it provides broad applicability without any labeling, indicating that the intrinsic nature can be assessed. In the future it will offer a less expensive and faster method with many practical uses, maintaining high sensitivity, stability and specificity. Especially, disposable electrodes can be fabricated easily through screen printed electrodes. Most of impedance spectroscopy has been carried out by frequency response analyzers, where the measurement has been done in the frequency domain. Recently, several novel techniques have been reported such as a complemental measurement in the time domain for high speed acquisition [49–51], adaptive filtering technique [52], which is highly expected to accelerate further advanced research in real-time impedance spectroscopy.

References

1. Debye, P. (1929). Polar molecules. *J Soc Chem Ind* 48 (43): 1036–1037.
2. Cole, K.S. and Cole, R.H. (1941). Dispersion and absorption in dielectrics I. Alternating current characteristics. *J Chem Phys* 9 (4): 341–351.
3. Manning, M.F. and Bell, M.E. (1940). Electrical conduction and related phenomena in solid dielectrics. *Rev Mod Phys* 12 (3): 215–256.
4. Havriliak, S. and Negami, S. (1967). A complex plane representation of dielectric and mechanical relaxation processes in some polymers. *Polymer* 8: 161–210.
5. Warburg, E. (1899). Ueber die Spitzenentladung. *Ann Phys Chem* 303 (1): 69–83.
6. Berggren, C., Stålhandske, P., Brundell, J., and Johansson, G. (1999). A feasibility study of a capacitive biosensor for direct detection of DNA hybridization. *Electroanalysis* 11 (3): 156–160.

7 Sakong, J., Roh, H., and Roh, Y. (2007). Surface acoustic wave DNA sensor with micro-fluidic channels. *Jpn J Appl Phys* 46 (7B): 4729–4733.

8 Zerrouki, C., Fourati, N., Lucas, R., Vergnaud, J., Fougnion, J.-M., Zerrouki, R. et al. (2010). Biological investigation using a shear horizontal surface acoustic wave sensor: Small "Click generated" DNA hybridization detection. *Biosens Bioelectron* 26 (4): 1759–1762.

9 Kruchinin, A.A. and Vlasov, Y.G. (1996). Surface plasmon resonance monitoring by means of polarization state measurement in reflected light as the basis of a DNA-probe biosensor. *Sens Actuators B Chem* 30 (1): 77–80.

10 Peterlinz, K.A., Georgiadis, R.M., Herne, T.M., and Tarlov, M.J. (1997). Observation of hybridization and dehybridization of thiol-tethered DNA using two-color surface plasmon resonance spectroscopy. *J Am Chem Soc* 119 (14): 3401–3402.

11 Andle, J.C., Vetelino, J.F., Lade, M.W., and McAllister, D.J. (1992). An acoustic plate mode biosensor. *Sens Actuators B Chem* 8 (2): 191–198.

12 Berggren, C., Bjarnason, B., and Johansson, G. (2001). Capacitive biosensors. *Electroanalysis* 13 (3): 173–180.

13 Zhu, N., Gao, H., Gu, Y., Xu, Q., He, P., and Fang, Y. (2009). PAMAM dendrimer-enhanced DNA biosensors based on electrochemical impedance spectroscopy. *Analyst* 134 (5): 860.

14 Gebala, M., Stoica, L., Neugebauer, S., and Schuhmann, W. (2009). Label-free detection of DNA hybridization in presence of intercalators using electrochemical impedance spectroscopy. *Electroanalysis* 21 (3–5): 325–331.

15 Sha, F. and Chen, F.-M. (2000). Actinomycin D binds strongly to d(CGACGACG) and d(CGTCGTCG). *Biophys J* 79 (4): 2095–2104.

16 Borrego-Varillas, R., Nenov, A., Kabaciński, P., Conti, I., Ganzer, L., Oriana, A. et al. (2021). Tracking excited state decay mechanisms of pyrimidine nucleosides in real time. *Nat Commun* 12 (1): 7285.

17 Pecourt, J.-M.L., Peon, J., and Kohler, B. (2000). Ultrafast internal conversion of electronically excited RNA and DNA nucleosides in water. *J Am Chem Soc* 122 (38): 9348–9349.

18 Prokhorenko, V.I., Picchiotti, A., Pola, M., Dijkstra, A.G., and Miller, R.J.D. (2016). New insights into the photophysics of DNA nucleobases. *J Phys Chem Lett* 7 (22): 4445–4450.

19 Buchner, F., Nakayama, A., Yamazaki, S., Ritze, H.-H., and Lübcke, A. (2015). Excited-state relaxation of hydrated thymine and thymidine measured by liquid-jet photoelectron spectroscopy: Experiment and simulation. *J Am Chem Soc* 137 (8): 2931–2938.

20 Hlavata, L., Benikova, K., Vyskocil, V., and Labuda, J. (2012). Evaluation of damage to DNA induced by UV-C radiation and chemical agents using electrochemical biosensor based on low molecular weight DNA and screen-printed carbon electrode. *Electrochimica Acta* 71: 134–139.

21 Ziyatdinova, G. and Labuda, J. (2011). Complex electrochemical and impedimetric evaluation of DNA damage by using DNA biosensor based on a carbon screen-printed electrode. *Anal Methods* 3 (12): 2777.

22 Qiu, Y., Qu, X., Dong, J., Ai, S., and Han, R. (2011). Electrochemical detection of DNA damage induced by acrylamide and its metabolite at the graphene-ionic liquid-Nafion modified pyrolytic graphite electrode. *J Hazard Mater* 190 (1–3): 480–485.

23 Ensafi, A.A., Kazemnadi, N., Amini, M., and Rezaei, B. (2015). Impedimetric DNA-biosensor for the study of dopamine induces DNA damage and investigation of inhibitory and repair effects of some antioxidants. *Bioelectrochemistry* 104: 71–78.

24 Gong, Q., Han, H., Yang, H., Zhang, M., Sun, X., Liang, Y. et al. (2019). Sensitive electrochemical DNA sensor for the detection of HIV based on a polyaniline/graphene nanocomposite. *J Materiomics* 5 (2): 313–319.

25 Huo, H., He, Y., Chen, W., Wu, L., Yi, X., and Wang, J. (2022). Simultaneously monitoring UVC-induced DNA damage and photoenzymatic repair of cyclobutane pyrimidine dimers by electrochemical impedance spectroscopy. *Talanta* 239: 123081.

26 Sipe, J.D., Benson, M.D., Buxbaum, J.N., Ikeda, S., Merlini, G., Saraiva, M.J.M. et al. (2016). Amyloid fibril proteins and amyloidosis: Chemical identification and clinical classification International Society of Amyloidosis 2016 Nomenclature Guidelines. *Amyloid* 23 (4): 209–213.

27 Nelson, R., Sawaya, M.R., Balbirnie, M., Madsen, A.Ø., Riekel, C., Grothe, R. et al. (2005). Structure of the cross-β spine of amyloid-like fibrils. *Nature* 435 (7043): 773–778.

28 Jaroniec, C.P., MacPhee, C.E., Bajaj, V.S., McMahon, M.T., Dobson, C.M., and Griffin, R.G. (2004). High-resolution molecular structure of a peptide in an amyloid fibril determined by magic angle spinning NMR spectroscopy. *Proc Natl Acad Sci USA* 101 (3): 711–716.

29 Eisenberg, D. and Jucker, M. (2012). The amyloid state of proteins in human diseases. *Cell* 148 (6): 1188–1203.

30 Fitzpatrick, A.W.P., Debelouchina, G.T., Bayro, M.J., Clare, D.K., Caporini, M.A., Bajaj, V.S. et al. (2013). Atomic structure and hierarchical assembly of a cross-β amyloid fibril. *Proc Natl Acad Sci USA* 110 (14): 5468–5473.

31 Petkova, A.T., Leapman, R.D., Guo, Z., Yau, W.-M., Mattson, M.P., and Tycko, R. (2005). Self-propagating, molecular-level polymorphism in Alzheimer's ß-amyloid fibrils. *Science* 307 (5707): 262–265.

32 Goldsbury, C.S., Cooper, G.J.S., Goldie, K.N., Müller, S.A., Saafi, E.L., Gruijters, W.T.M. et al. (1997). Polymorphic fibrillar assembly of human amylin. *J Struct Biol* 119 (1): 17–27.

33 Wälti, M.A., Ravotti, F., Arai, H., Glabe, C.G., Wall, J.S., Böckmann, A. et al. (2016). Atomic-resolution structure of a disease-relevant Aβ(1–42) amyloid fibril.

Proc Natl Acad Sci USA [Internet] [cited 2022 Mar 27] 113 (34). https://pnas.org/doi/full/10.1073/pnas.1600749113.

34 Colvin, M.T., Silvers, R., Ni, Q.Z., Can, T.V., Sergeyev, I., Rosay, M. et al. (2016). Atomic resolution structure of monomorphic Aβ_{42} amyloid fibrils. *J Am Chem Soc* 138 (30): 9663–9674.

35 Bunce, S.J., Wang, Y., Stewart, K.L., Ashcroft, A.E., Radford, S.E., Hall, C.K. et al. (2019). Molecular insights into the surface-catalyzed secondary nucleation of amyloid-β_{40} (Aβ_{40}) by the peptide fragment Aβ_{16-22}. *Sci Adv* 5 (6): eaav8216.

36 Biancalana, M. and Koide, S. (2010). Molecular mechanism of Thioflavin-T binding to amyloid fibrils. *Biochim Biophys Acta Proteins Proteom* 1804 (7): 1405–1412.

37 Turnbull, J.H., Weiss, J., Butler, J.A.V., Szent-Győrgy, A., Chance, B., Baughan, E.C. et al. (1959). General discussion. *Discuss Faraday Soc* 27 (0): 232–272.

38 Morgan, H. and Pethig, R. (1986). Protonic and ionic conduction in lysozyme. Hydration and field-dependent effects. *J Chem Soc, Faraday Trans 1* 82 (1): 143.

39 Awasthi, P., Singh, A., Khatun, S., Gupta, A.N., and Das, S. (2020). Fibril growth captured by electrical properties of amyloid- β and human islet amyloid polypeptide. *Phys Rev E* 101 (6): 062413.

40 Shipps, C., Kelly, H.R., Dahl, P.J., Yi, S.M., Vu, D., Boyer, D. et al. (2021). Intrinsic electronic conductivity of individual atomically resolved amyloid crystals reveals micrometer-long hole hopping via tyrosines. *Proc Natl Acad Sci USA* 118 (2): e2014139118.

41 da Silva, R.R., de Lima, S.V., de Oliveira, H.P., de Melo, C.P., Frías, I.A.M., Oliveira, M.D.L. et al. (2017). Real-time monitoring of amyloid fibrillation by electrical impedance spectroscopy. *Colloids Surf B Biointerfaces* 160: 724–731.

42 Groenning, M., Norrman, M., Flink, J.M., van de Weert, M., Bukrinsky, J.T., Schluckebier, G. et al. (2007). Binding mode of Thioflavin T in insulin amyloid fibrils. *J Struct Biol* 159 (3): 483–497.

43 Rushworth, J.V., Ahmed, A., Griffiths, H.H., Pollock, N.M., Hooper, N.M., and Millner, P.A. (2014). A label-free electrical impedimetric biosensor for the specific detection of Alzheimer's amyloid-beta oligomers. *Biosens Bioelectron* 56: 83–90.

44 Balkenhohl, T. and Lisdat, F. (2007). Screen-printed electrodes as impedimetric immunosensors for the detection of anti-transglutaminase antibodies in human sera. *Anal Chim Acta* 597 (1): 50–57.

45 Rohrbach, F., Karadeniz, H., Erdem, A., Famulok, M., and Mayer, G. (2012). Label-free impedimetric aptasensor for lysozyme detection based on carbon nanotube-modified screen-printed electrodes. *Anal Biochem* 421 (2): 454–459.

46 Quintana, J.C., Arduini, F., Amine, A., Punzo, F., Destri, G.L., Bianchini, C. et al. (2011). Part I: A comparative study of bismuth-modified screen-printed electrodes for lead detection. *Anal Chim Acta* 707 (1–2): 171–177.

47 Tran, L.D., Nguyen, B.H., Van Hieu, N., Tran, H.V., Nguyen, H.L., and Nguyen, P.X. (2011). Electrochemical detection of short HIV sequences on chitosan/ Fe3O4 nanoparticle based screen printed electrodes. *Mater Sci Eng C* 31 (2): 477–485.

48 Escamilla-Gómez, V., Campuzano, S., Pedrero, M., and Pingarrón, J.M. (2009). Gold screen-printed-based impedimetric immunobiosensors for direct and sensitive Escherichia coli quantisation. *Biosens Bioelectron* 24 (11): 3365–3371.

49 Park, S.-M., Yoo, J.-S., Chang, B.-Y., and Ahn, E.-S. (2006). Novel instrumentation in electrochemical impedance spectroscopy and a full description of an electrochemical system. *Pure Appl Chem* 78 (5): 1069–1080.

50 Yoo, J.-S. and Park, S.-M. (2000). An electrochemical impedance measurement technique employing fourier transform. *Anal Chem* 72 (9): 2035–2041.

51 Chang, B.-Y. and Park, S.-M. (2007). Fourier transform analysis of chronoamperometric currents obtained during staircase voltammetric experiments. *Anal Chem* 79 (13): 4892–4899.

52 Stupin, D.D., Koniakhin, S.V., Verlov, N.A., and Dubina, M.V. (2017). Adaptive filtering to enhance noise immunity of impedance and admittance spectroscopy: Comparison with fourier transformation. *Phys Rev Applied* 7 (5): 054024.

5

Mass Spectrometry Imaging

Shuichi Shimma

Department of Engineering, Osaka University, Suita, Osaka, Japan

5.1 Introduction

In mass spectrometry (MS), the molecules in the sample are ionized using various methods. Differences in the motion of the resulting ions in an electromagnetic field are used to determine the mass (actually a dimensionless quantity called m/z, mass divided by charge) on the horizontal axis and the intensity of the detected ions on the vertical axis, providing a graphical data set called a "mass spectrum," a component analysis technique.

Mass spectrometry imaging (MSI), the subject of this chapter, is a technique for performing mass spectrometry directly on the surface of a sample and visualizing the distribution on the sample surface from the obtained mass spectrum. When biomolecules are analyzed using MS, the sample is usually ground, extracted, and purified, followed by qualitative and quantitative analysis. The existing methods can thus reveal "how many molecules are contained in a sample." However, detailed information regarding molecular distribution is lost after extraction, and it is difficult to provide highly accurate distribution information to answer the questions of "where (spatial information), what (qualitative), and how much (quantitative)."

MSI is a new imaging technique reported in 1997 [1]. At that time, MSI was one of new techniques for protein imaging [2, 3]. Various sample preparation methods have been proposed for protein imaging [4], and many researchers joined the trend and initiated MSI-based research. As the number of MSI researchers increased, the targets of measurement expanded from proteins to phospholipids [5, 6] and metabolites [7], and drug measurement became possible

Analytical Techniques for the Elucidation of Protein Function, First Edition. Edited by Isao Suetake, Rohit K. Sharma, and Hironobu Hojo.

[8, 9]. New measurement methods such as structural isomer separation imaging and enzyme activity imaging have also been reported [10–14] and are described in this chapter.

Overall, this chapter describes the basic knowledge of MSI and its applications. In addition to protein imaging, the applications of MSI include structural isomer imaging and enzyme activity imaging to visualize enzyme function as described above.

5.2 Workflow of MSI

The workflow of MSI is shown in Figure 5.1. Unlike other imaging methods such as positron emission tomography, MSI directly detects target molecules on the surface of tissue sections prepared from collected tissues using MS. Therefore, sample collection is necessary for MSI. For example, human samples are collected from surgical specimens or biopsies, and animal samples are collected from rodents to obtain organs or tumor tissues. The collected samples are generally frozen using liquid nitrogen. When freezing, MSI requires fresh freezing rather than chemical fixation. Frozen samples are then sectioned to 10–20 μm thickness using a cryomicrotome and placed on a conductive transparent glass plate. The glass plate used here is usually made of indium-tin-oxide (ITO) glass.

Matrix-assisted laser desorption/ionization (MALDI), involving laser irradiation of the sample, is generally used as the ionization method in MSI. Here, matrices play the role of supporting ionization. After irradiation, matrices absorb the laser energy and then sublimate (solid state to gas state) the sample molecules. During this process, a single charge is normally provided to molecules. After the tissue sections are surface coated with a matrix as an ionization aid, they are introduced into the instrument, and a data point, or laser irradiation location, is set on the sample surface. Mass spectra are then automatically acquired at each point and stored along with positional information. The ion intensities of the peaks of interest are extracted from the obtained mass spectrum, and intensity distributions are generated to obtain the imaging results. The interval between the data points set here corresponds to the resolution of MSI.

Based on the above measurement principle, MSI is characterized by its ability to draw distributions for all the peaks ionized and detected on the sample surface. Thus, MSI allows researchers to obtain distribution information for various molecules simultaneously without requiring the use of tags (fluorescence, radiation, and so on) for visualization. Tandem mass spectrometry (MS/MS or MS^n) can also be used for obtaining structural information on the molecules in mass spectrometry. Using MS^n on tissues, structural information can be obtained by narrowing down the measurement target, and imaging data can be obtained with very high specificity.

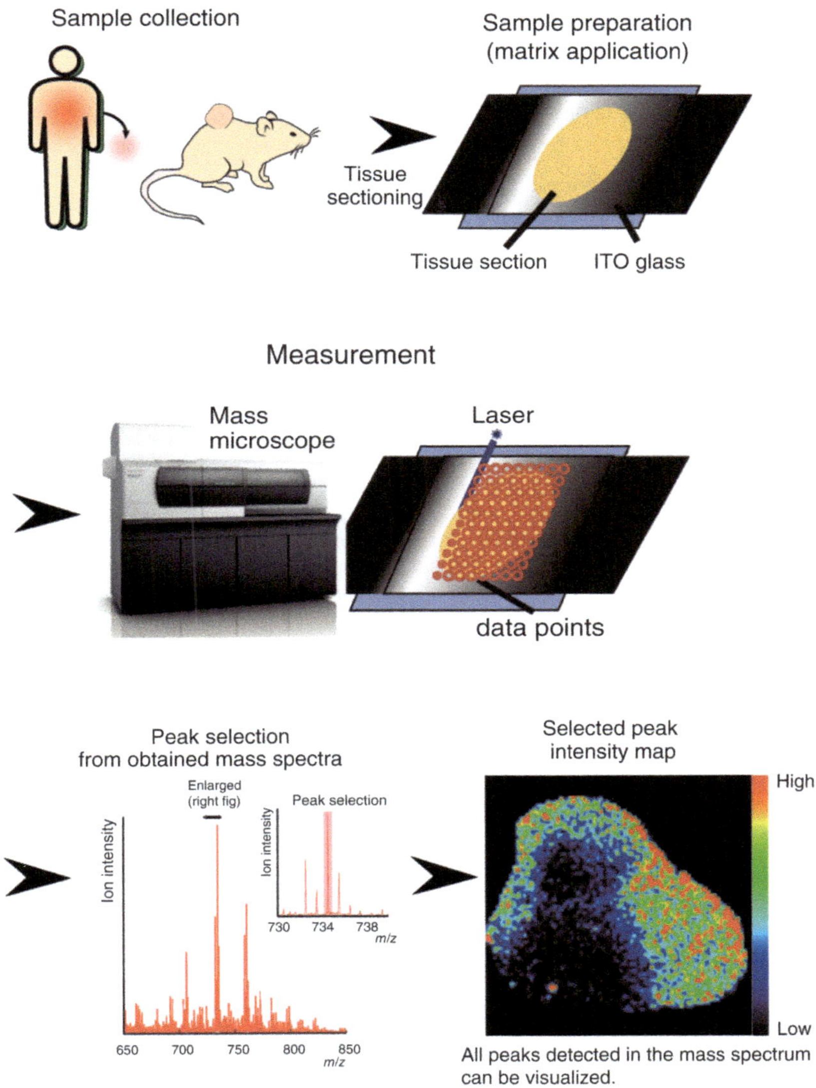

Figure 5.1 Mass spectrometry imaging workflow. In MSI, collected tissue is frozen, sections are prepared, and a matrix (an ionization supportive compound) is applied to the surface. After direct mass analysis of the tissue surface, the peak values of *m/z* to be imaged are extracted, and distribution information is obtained as peak intensity maps. Reproduced with permission of Shimadzu Scientific Instruments.

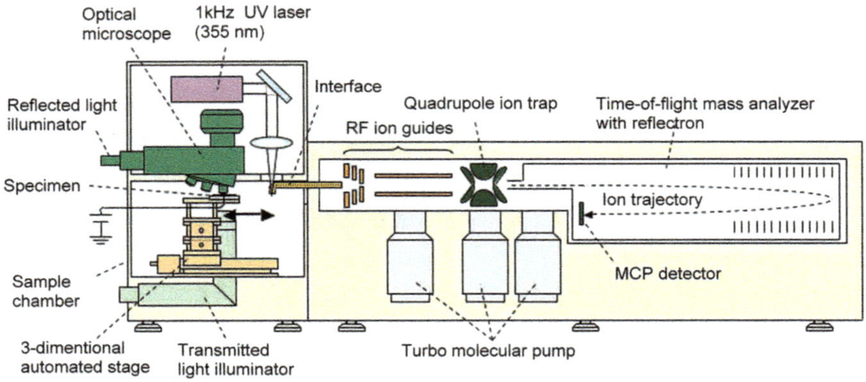

Figure 5.2 Schematic drawing of a mass microscope. The left chamber is maintained at atmospheric pressure and houses the microscope, sample stage, and laser for MALDI. The chamber on the right houses a tandem mass spectrometer with an ion trap and time-of-flight under vacuum. The left and right chambers are connected by a differential pumping system between atmospheric pressure and high vacuum. Adapted with permission from Harada et al. [15]. Copyright 2007 American Chemical Society. Harada T, Yuba-Kubo A, Sugiura Y, Zaima N, Hayasaka T, Goto-Inoue N, et al. "Visualization of volatile substances in different organelles with an atmospheric-pressure mass microscope." *Anal Chem.* 2009; 81(21): 9153–7.

5.3 Mass Microscope

Although various mass spectrometers can be used to perform MSI, a mass microscope is described here. The mass microscope is a Japan-original MSI-specific instrument, which facilitates microscopic observation of samples under atmospheric pressure using atmospheric pressure MALDI as an ion source (Figure 5.2) [15]. The mass microscope thus provides very detailed molecular distribution information by MSI. The mass spectrometer part of the mass microscope is a tandem mass spectrometer equipped with an ion trap, which can acquire product ion spectra with high sensitivity, and a time-of-flight mass spectrometer with high mass resolution. Inside the ion source is an *XY* movable stage, a microscope, and a variable-diameter Nd:YAG laser with a repetition rate of 1 kHz for ionization (minimum laser diameter: 5 μm). Using this system, small samples such as needle biopsies (1.2 mm × 0.6 mm) can be imaged with 40 μm resolution in approximately 30 minutes, including sample preparation.

5.4 Visualization of Small Molecules (Pharmaceutical)

In recent years, anti-cancer drugs have mainly been molecularly targeted drugs that act on specific targets in cancer cells. Application of MSI for the pharmacokinetic evaluation of molecularly targeted drugs has attracted significant attention because it enables visualization of the mechanism of action in vivo after drug administration without the need for labeling. Figure 5.3 shows an example of MSI of a PARP inhibitor (drug name: olaparib) in a mouse tumor model tissue derived from human triple-negative breast cancer cells [16]. Tumor cells were implanted subcutaneously in nude mice (BALB/c), and tumor tissues were obtained under isoflurane anesthesia, 30 minutes after 50 mg/kg olaparib was administered intraperitoneally to the tumor model mice. Tumors were frozen at −80 °C in liquid nitrogen without chemical fixation. Frozen sections

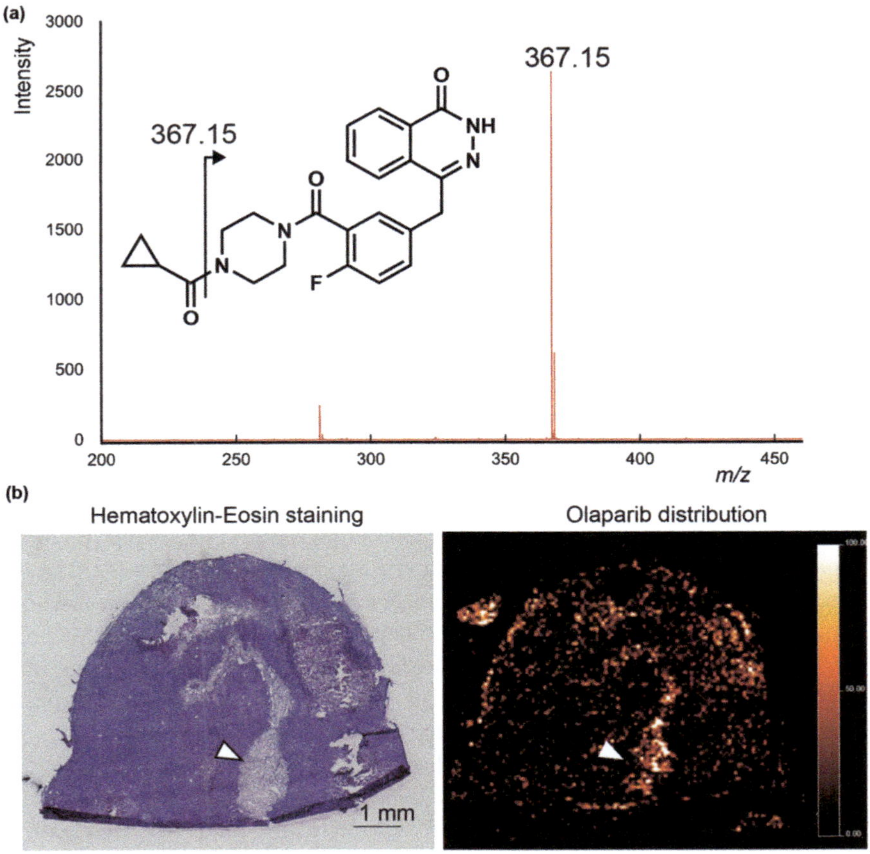

Figure 5.3 An example of MSI for a pharmaceutical product. (a) Product ion spectrum of Olaparib. The peak at *m/z* = 367.15, derived from olaparib, is clearly detected. (b) Olaparib distribution in mouse subcutaneous tumors (breast cancer-derived cell line). Left: Hematoxylin-eosin staining results; Right: Olaparib distribution. An accumulation site is observed in olaparib distribution, but the HE staining results indicate a necrotic area (arrowhead). Scale bar: 1 mm. [16], Shimma et al. (2013) / Adapted with permission from John Wiley & Sons.

were prepared at 8 μm thickness, followed by matrix application at room temperature; product ion spectra were then acquired at 35 μm data intervals using the mass microscope.

Figure 5.3(a) shows the obtained product ion spectrum. The peak at $m/z = 367.15$ derived from the olaparib structure could be clearly detected. The peak intensity distribution of $m/z = 367.15$ (Figure 5.3b) confirmed that the signal was distributed throughout the tumor tissue, but localization of the signal was also observed. Hematoxylin-eosin staining of the samples after measurement revealed that the signal was localized in the necrotic area. Thus, MSI can provide detailed distributional information of drugs by mapping tissue staining images to ion distribution information on a one-to-one basis. Similar to olaparib, as described here, MSI can be used to evaluate the pharmacokinetics of various other anti-cancer drugs (molecularly targeted drugs and cytotoxic drugs) in animal models and human clinical specimens.

5.5 Structural Isomer Discrimination Imaging (Steroid Hormones)

MSI has recently attracted attention for its ability to detect and visualize various molecules directly. It can thus be used for steroid hormones, which have a steroid skeleton and are essential molecules with a variety of physiological regulatory functions, in vivo.

When imaging steroid hormones using conventional MALDI, their low polarity results in a very low ionization efficiency. "On-tissue derivatization" is a sample pretreatment method that solves this problem and provides higher sensitivity. Derivatization is a well-known sample preparation method in analytical chemistry; simply spraying a reagent can achieve high sensitivity, and in some cases, molecules that were completely undetectable can be detected using this method. Figure 5.4 shows the results of a tissue derivatization method using Girard's reagent T (GirT) and pyridine sulfur trioxide complex, respectively. Using MS/MS, the ions dissociated by binding of the introduced functional groups, $m/z = 343.15$ (GT-Ts) and $m/z = 96.96$, 269.21 (TsS), can be detected [17].

This new sample preparation method was applied to a sample from a patient with primary aldosteronism (PA), a metabolic disorder, to visualize aldosterone. PA is a form of secondary hypertension that accounts for about 10% of hypertensive patients (43 million in Japan). In PA, autonomous overproduction of adrenal aldosterone causes hypertension. Aldosterone overproduction is known to be caused by aldosterone-producing adenomas (APA) in the adrenal glands. Immunostaining of enzymes involved in aldosterone production was reported in 2010 [18], and various pathological conditions have been observed since, but no

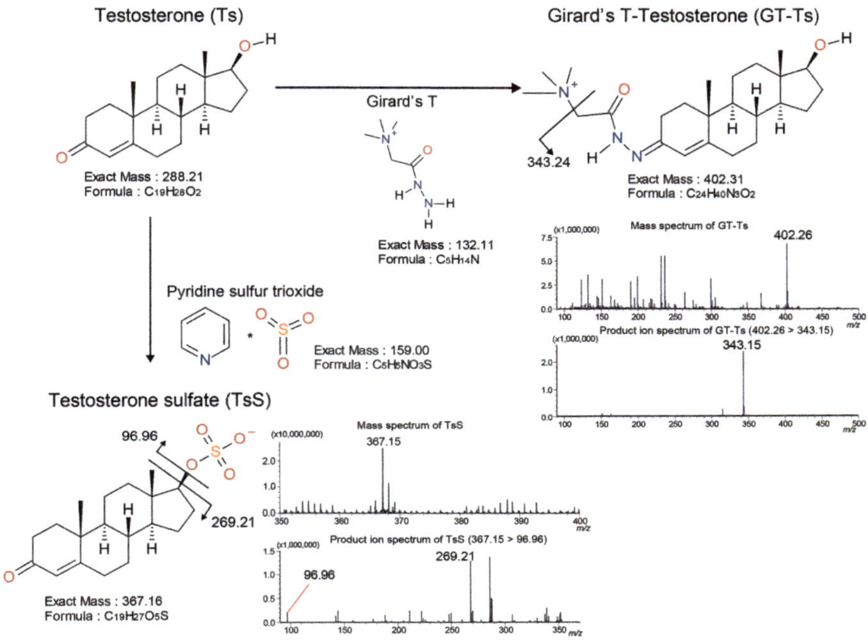

Figure 5.4 Overview of reagents for on-tissue derivatization and the mass spectra obtained from each derivatized standard testosterone sample. Positive ion detection for GT-Ts (GirT derivative of testosterone) and negative ion detection for TsS (pyridine trioxyde derivative of testosterone) were used. In the TsS spectra, the precursor ion at m/z = 367.15 and two different fragment ions (m/z = 96.96 and 269.12) were detected. In the GT-Ts spectra, the precursor ion at m/z = 402.26 and the fragment ion at m/z = 343.15 with low background were detected. Reproduced with permission from [17] / Springer Nature.

method has been developed to visualize the localization of aldosterone itself. Therefore, the sites of aldosterone overproduction have not been identified, and the pathogenesis of the disease remains to be elucidated.

As already explained, MALDI-MSI can be used to directly visualize the localization of biomolecules in vivo. However, there are two problems with aldosterone: (i) the total amount of aldosterone and its ionization efficiency are low; (ii) further, as shown in Figure 5.5(a), a structural isomer (cortisone) of aldosterone exists in the adrenal gland. Therefore, to solve (i), the above-mentioned tissue derivatization method using GirT was applied. Figure 5.5(b) shows the MS/MS spectrum of m/z = 474.3 obtained as a derivative from aldosterone and cortisone standards. This allows detection of m/z = 415.2, from which the trimethylamine is desorbed. However, the problem of structural isomer separation shown in (ii) has not been solved because m/z = 415.2 was detected in both samples. Therefore, even if MS/

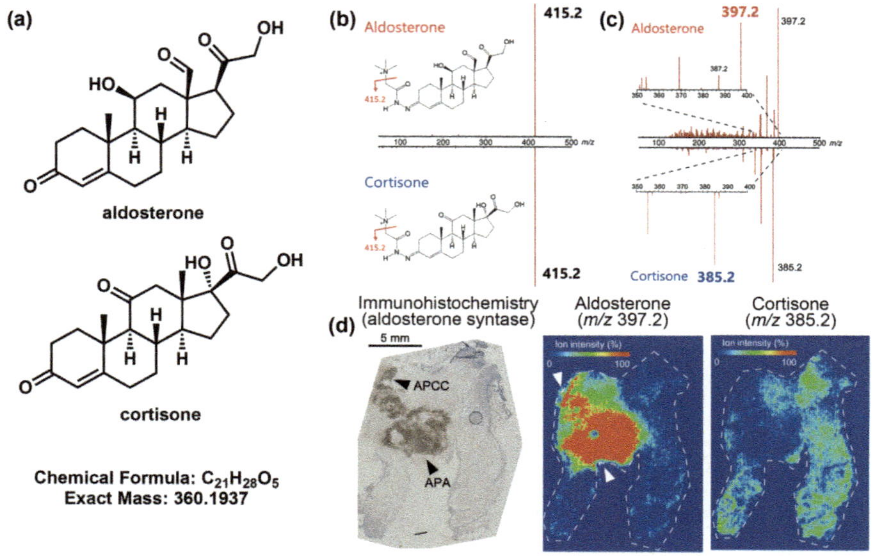

Figure 5.5 Structural isomer separation in steroid hormone MSI. (a) Structures of aldosterone and cortisone. As both are $C_{21}H_{28}O_5$, they cannot be separated by mass spectrometry. (b) Product ion spectrum of aldosterone and cortisone after GirT derivatization. The peak *m/z* at 415.2, a fragment peak of GirT, is detected and cannot be separated. (c) Product ion spectrum of *m/z* = 415.2. The *m/z* = 397.2 from aldosterone and *m/z* = 385.2 from cortisone are detected as specific peaks. (d) Aldosterone and cortisone separation imaging results using clinical samples. Aldosterone accumulates in aldosterone-producing adenomas (APA) and aldosterone-producing cell clusters (APCC), and cortisone is nested with aldosterone. [13], Takeo et al. (2019) / Adapted with permission from American Chemical Society.

MS is performed on tissue, it is impossible to separate aldosterone from cortisone, and it is difficult to specifically detect and visualize the aldosterone distribution in the adrenal gland.

To solve this problem, a product ion spectrum was obtained from $m/z = 415.2$ as a precursor ion, as shown in Figure 5.5(c). In the product ion spectrum obtained, specific peaks were derived from aldosterone ($m/z = 397.2$) and cortisone ($m/z = 385.2$). Thus, it is possible to improve the detection sensitivity and to distinguish structural isomers by performing tissue derivatization with GirT and MS^3; we thus developed a novel aldosterone-specific distribution visualization method using MALDI-MSI [13]. This method was applied to human PA adrenal specimens containing APA and aldosterone-producing cell clusters (APCC); both APA and APCC were identified as aldosterone overproduction sites by the specific and direct visualization of aldosterone. These results represent the first success story of MSI in the field of endocrinology (Figure 5.5d).

5.6 Visualization of Proteins (Intact, Digestion)

Currently, MSI is often used to target small molecule compounds. However, reports of intact protein imaging have also been found in recent years [19–23]. For example, Paulini et al. [22] performed a proteomic study of bovine ovarian tissue and revealed differences in the molecular profiles of the precursor follicles as well as the spatial distribution of the most abundant molecular components in this tissue. However, in the mass spectra of proteins, which are macromolecules, the peaks are broad, and it is challenging to identify the protein that originated them. Thus, MSI of the protein itself remains difficult in terms of molecular identification and detection sensitivity. To solve this problem, there is a method involving denaturation of the proteins on the sample tissue, followed by the detection of enzymatically digested peptides. This is called the on-tissue digestion method, and various pretreatment protocols have been introduced for it [24, 25].

Figure 5.6 shows an example of the MSI of digested peptides obtained by on-tissue digestion in mouse cerebellar sections. Product ion spectra were obtained on the tissue sections, and the amino acid sequences were identified. In this mass spectrum, several peaks were obtained despite the high background, and among these, the distributions of $m/z = 1460$ and $m/z = 1743$ were obtained in particular (Figure 5.6c). The obtained distribution shows that the two are in a nested relationship. As shown in Figures 5.6(f) and 5.6(g), $m/z = 1460$ is derived from myelin basic protein (MBP), and $m/z = 1743$ is from histone H2B F. Thus, it is possible to detect proteins as digested peptides using on-tissue digestion and to estimate their amino acid sequences using MS/MS.

This technique has been applied in research on cancer and neurodegenerative diseases. Cordeiro et al. [26] performed MSI using on-tissue digestion to search for biomarkers in tumor tissues; they found FNDC1, A1BG, PDIA3, HSPA5, and calnexin as important prognostic markers in patients with breast cancer. As exemplified by this study, on-tissue digestion has also been used with formalin-fixed, paraffin-embedded (FFPE) blocks for protein imaging. Karayel-Basar et al. [27] also used on-tissue digestion combined with liquid chromatography-mass spectrometry on samples from Huntington's disease transgenic and wild-type mice to analyze the differences in protein-derived signals. They reported that 22 proteins are differentially expressed.

However, this method also has its problems. Owing to the nonspecific digestion of tissue surface proteins, mainly by enzymatic digestion using trypsin, many of the detected signals are derived from cytoskeletal proteins, which hinders the detection of trace proteins. Therefore, to make protein MSI practical, further technological development, including preprocessing, is necessary for detecting the target proteins from a large number of protein signals.

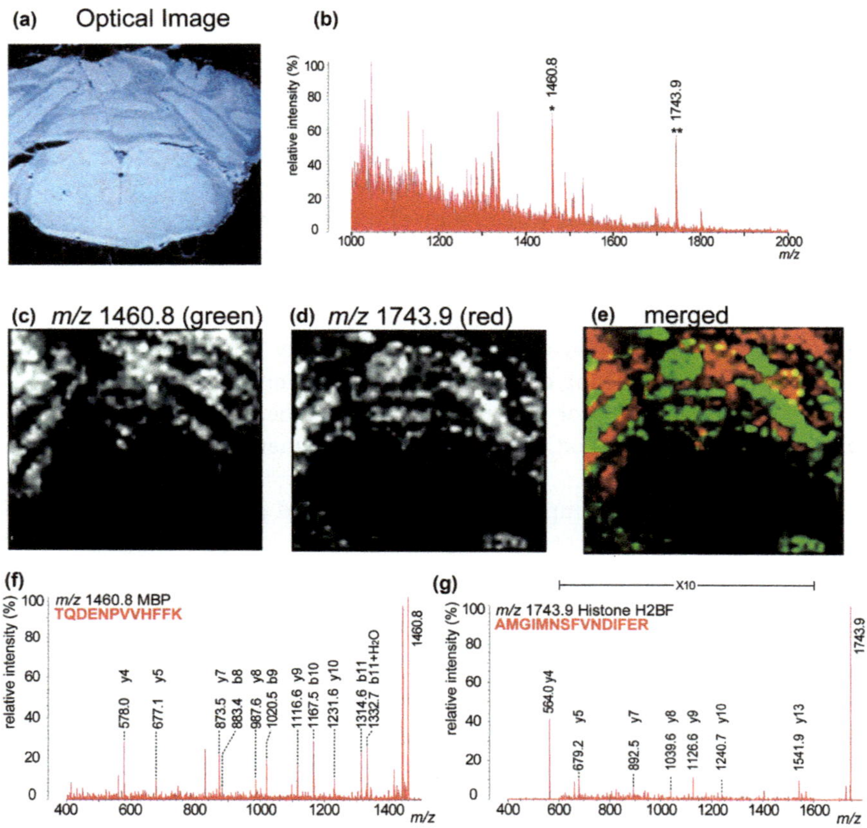

Figure 5.6 Example of tryptic-digested protein imaging and precursor ion mass spectra using the positive ion detection mode. (a) Optical image of the imaging region, and (b) accumulated mass spectrum from the imaging region. Images (c) and (d) show the imaging results of m/z = 1460.8 and 1743.9, which are labeled by asterisks in (b). The merged image (red, m/z = 1460.8; and green, m/z = 1743.9) is shown in (e). These peaks were identified by direct MS^n and were identified as fragment ions of myelin basic protein (MBP) (f) and histone H2B (g). [25], Shimma et al. (2008) / Adapted with permission from American Chemical Society.

5.7 Visualization of Protein Function (Enzymatic Activity Visualization)

Immunohistochemistry (IHC) is a commonly used method to determine enzyme distribution in tissues. However, IHC can visualize the presence of enzymes but not their activity. A technique called enzyme histochemistry has long been known as a method to visualize the distribution of enzyme activity that cannot be obtained by IHC [28]. In enzyme histochemistry, a substrate is supplied to the surface of a sample and is visualized by a color reaction through an endogenous enzymatic reaction. In conventional enzyme histochemistry, the color reaction step is not versatile as the steps of the color reaction need to be designed specifically for each case. Here, we introduce a method to visualize molecules produced in enzymatic reactions by MSI to solve these problems.

The activity distribution of acetylcholine esterase (AChE) and choline acetyltransferase (ChAT) as target enzymes in tissues has been demonstrated [11, 12]. AChE is an enzyme that uses acetylcholine as a substrate and degrades it into choline and acetic acid. In contrast, ChAT is an enzyme that synthesizes acetylcholine from acetyl-CoA and choline as the reverse reaction (Figure 5.7a). Therefore, when detecting AChE activity, acetylcholine is supplied to the sample surface, and the degradation product, choline, is detected (Figure 5.7b); in the case of ChAT, the reverse reaction is used wherein choline and acetyl-CoA are supplied, and acetylcholine is detected. However, the supplied reagents must be distinguished from endogenous choline and acetylcholine. Therefore, deuterium-labeled acetylcholine-d9 and choline-d9 are used as substrates to detect choline-d9 and acetylcholine-d9 obtained in the enzymatic reaction. When imaging ChAT activity, the synthesized acetylcholine-d9 is quickly degraded by AChE. The cholinesterase inhibitor physostigmine is supplied before the substrate to prevent the

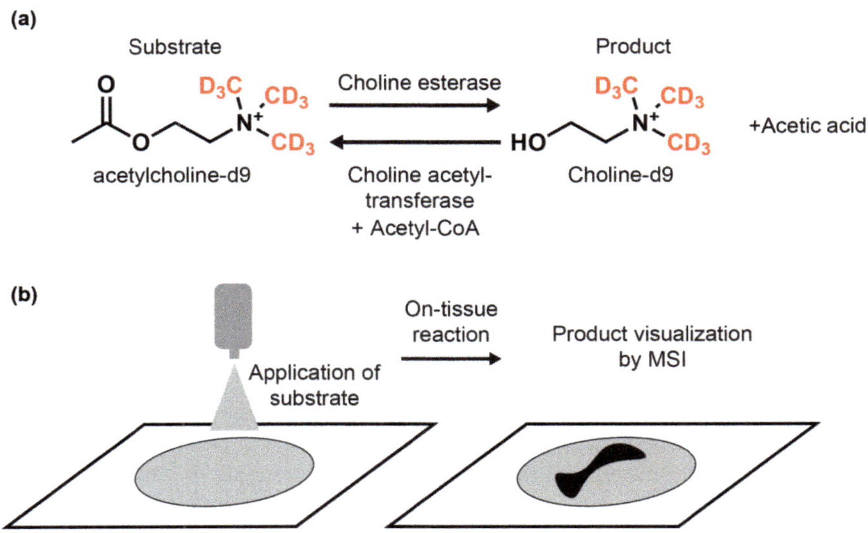

Figure 5.7 Concept of enzyme histochemistry using MSI. (a) Schematic drawing of decomposition and synthesis for acetylcholine and choline. (b) Workflow of MSI-based enzyme histochemistry. Adapted from [11].

degradation of synthesized acetylcholine-d9. All reagents, including substrates, can be supplied by manual spraying during sample preparation. The matrix used is α-cyano-4-hydroxycinnamic acid (α-CHCA, Merck, Darmstadt, Germany) and is supplied as a combination of vacuum deposition using iMLayer (Shimadzu, Kyoto, Japan) and spraying α-CHCA solution [16].

In determining the relative enzyme activity distribution, a difference in ionization efficiency between choline-d9 and acetylcholine-d9 was confirmed and used for correction in later calculations. As shown in Figure 5.8(a), the ionization efficiency of acetylcholine-d9 is three times higher than that of choline-d9. The final relative enzyme activity distribution was thus reconstructed by image calculation using the following formula from the obtained distribution results of choline-d9 and acetylcholine-d9 (Figure 5.8b).

$$\textit{Relative ChE activity}(a.\ u.) = 100 \times \frac{3(I_{Ch-d9})}{(I_{ACh-d9}) + 3(I_{Ch-d9})}$$

The coefficient placed in front of $I_{Ch\text{-}d9}$ is the ionization efficiency of acetylcholine and choline, which was set to 3 based on the results of Figure 5.8(a). As shown in Figure 5.8(c), AChE activity was higher in the striatum (CPu), hippocampus (HP), and hypothalamus (TH) and was lower in the corpus callosum (CC) and cerebellar cortex (CBX). These results were comparable with those obtained using the existing enzyme histochemistry techniques, thus demonstrating the proof of concept of this method.

On the contrary, in the case of ChAT activity imaging, the obtained distribution results of choline-d9 and acetylcholine-d9 are calculated similarly, but the formula changes the molecule in the above equation to $I_{ACh\text{-}d9}$. Figure 5.9 shows the ChAT activity imaging results in mouse spinal cord using Luxol fast blue (LFB) and the IHC results for ChAT. Here, LFB is a commonly used staining method to visualize neural structures in the brain or spinal cord. The LFB results showed butterfly-shaped gray matter typical of spinal cord tissue and surrounding white matter. The immunostaining results showed positive brownish staining in the butterfly-shaped gray matter, with particularly strong staining in the ventral (anterior) root at the tissue periphery. In contrast, ChAT activity imaging results showed high activity in the anterior root and central gray matter. Interestingly, upon comparing the IHC results with those of ChAT activity distribution, the anterior root and anterior horn were not connected in IHC but were connected in the ChAT activity distribution (Figure 5.9 arrowheads). This might be attributed to the transmembrane and soluble forms of ChAT. In general, the staining process in IHC includes numerous washing steps, and soluble ChAT is thought to be washed away without being stained.

As the MSI method does not include a washing process, both transmembrane and soluble ChAT can be visualized. Thus, the ChAT connecting the anterior root

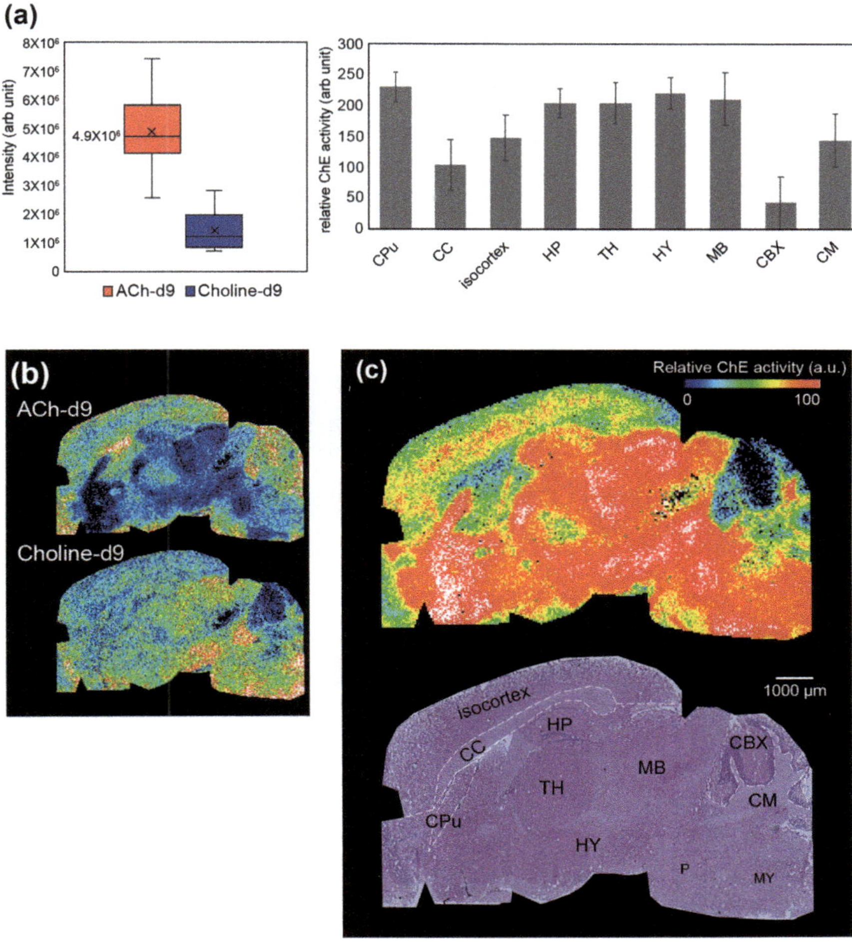

Figure 5.8 Evaluation of the normalization step. (a) Comparison of ionization efficiency between ACh-d9 and choline-d9 using mouse brain homogenates (n = 9). (b) Mass spectrometry images of ACh-d9 and choline-d9 in mouse whole brain sections (no pretreatment including normalization; lateral, 0.8 mm; pitch, 45 µm). (c) Cholinesterase (ChE) activity and hematoxylin-eosin (HE)-stained image after normalization. Heterogeneous ChE activity differed among different histological sites. CPu: striatum; CC: corpus callosum; HP: hippocampus; TH: thalamus; HY: hypothalamus; MB: midbrain; CBX: cerebellar cortex; CM: cerebellar medulla; P: pons; MY: medulla. (d) Relative comparison of ChE activity at each site. Scale bar: 1000 µm. [11], Takeo et al. (2020) / Adapted with permission from American Chemical Society.

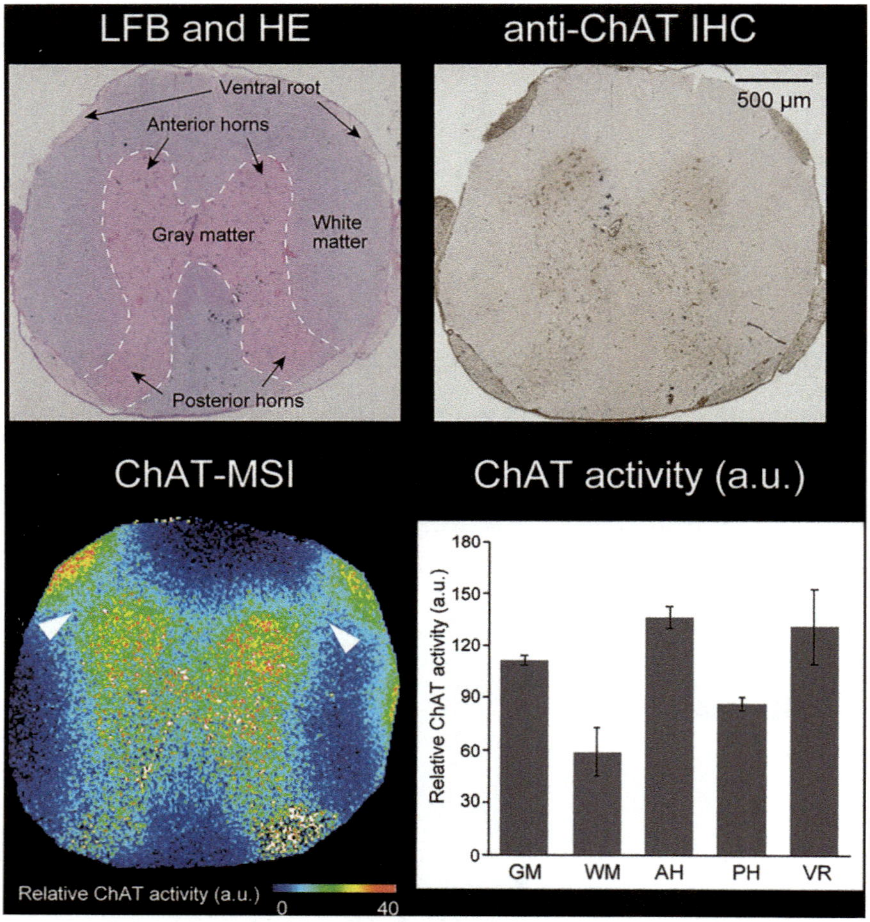

Figure 5.9 ChAT activity imaging in rat spinal cord tissue. Luxol fast blue (LFB) and hematoxylin-eosin (HE) staining, anti-ChAT immunohistochemistry, ChAT activity imaging by MSI at 15 μm pitch, and comparison of ChAT activity in different regions based on MSI results. Strong ChAT activity was observed in the anterior horn, ventral root, and ventral white matter (shown with white arrow) between the anterior horn and ventral root. Region of interest (ROI) analysis was performed using three sections from three rat spinal cords (n = 3). GM: gray matter, WM: white matter, AH: anterior horn, PH: posterior horn, VR: ventral root. Scale bar: 500 μm. [12], Takeo et al. (2021) / Adapted with permission from American Chemical Society.

and anterior horn may be soluble-type ChAT. As described above, enzyme activity visualization using MSI can facilitate understanding and observation of enzyme function and allow detection of proteins that cannot be observed using the existing staining methods.

5.8 Summary

Overall, this chapter introduced the basics and applications of MSI (visualization from small molecules to proteins). Currently, on-tissue digestion using digestive enzymes, such as trypsin, is the mainstay of protein MSI; however, MSI of intact proteins requires further technological development. A novel enzyme histochemical method to examine protein functions (enzyme activity) was also discussed. As the concept of enzyme histochemistry using MSI is very simple, it is expected to have various applications in the future when used in combination with IHC.

As shown in the examples introduced so far, MSI can now visualize various molecules. However, trial and error is required for sample preparation, such as matrix selection, to visualize new molecules. In this respect, MSI is a technology that still has room for further development.

References

1 Caprioli, R.M., Farmer, T.B., and Gile, J. (1997). Molecular imaging of biological samples: Localization of peptides and proteins using MALDI-TOF MS. *Anal Chem* 69 (23): 4751–4760.

2 Chaurand, P., Schwartz, S.A., and Caprioli, R.M. (2004). Assessing protein patterns in disease using imaging mass spectrometry. *J Proteome Res* 3 (2): 245–252.

3 Reyzer, M.L. and Caprioli, R.M. (2005). MALDI mass spectrometry for direct tissue analysis: A new tool for biomarker discovery. *J Proteome Res* 4 (4): 1138–1142.

4 Schwartz, S.A., Reyzer, M.L., and Caprioli, R.M. (2003). Direct tissue analysis using matrix-assisted laser desorption/ionization mass spectrometry: Practical aspects of sample preparation. *J Mass Spectrom* 38 (7): 699–708.

5 Good, C.J., Neumann, E.K., Butrico, C.E., Cassat, J.E., Caprioli, R.M., and Spraggins, J.M. (2022). High spatial resolution MALDI imaging mass spectrometry of fresh-frozen bone. *Anal Chem* 94 (7): 3165–3172.

6 Shimma, S., Sugiura, Y., Hayasaka, T., Hoshikawa, Y., Noda, T., and Setou, M. (2007). MALDI-based imaging mass spectrometry revealed abnormal distribution of phospholipids in colon cancer liver metastasis. *J Chromatogr B Analyt Technol Biomed Life Sci* 855 (1): 98–103.

7 Sugiyama, E., Skelly, A.N., Suematsu, M., and Sugiura, Y. (2020). In situ imaging of monoamine localization and dynamics. *Pharmacol Ther* 208: 107478.

8 Prideaux, B., Dartois, V., Staab, D., Weiner, D.M., Goh, A., Via, L.E. et al. (2011). High-sensitivity MALDI-MRM-MS imaging of moxifloxacin distribution in tuberculosis-infected rabbit lungs and granulomatous lesions. *Anal Chem* 83 (6): 2112–2118.

9 Prideaux, B. and Stoeckli, M. (2012). Mass spectrometry imaging for drug distribution studies. *J Proteomics* 75 (16): 4999–5013.

10 Hamilton, B.R., Marshall, D.L., Casewell, N.R., Harrison, R.A., Blanksby, S.J., and Undheim, E.A.B. (2020). Mapping enzyme activity on tissue by functional mass spectrometry imaging. *Angew Chem Int Ed Engl* 59 (10): 3855–3858.

11 Takeo, E., Fukusaki, E., and Shimma, S. (2020). Mass spectrometric enzyme histochemistry method developed for visualizing in situ cholinesterase activity in mus musculus and drosophila melanogaster. *Anal Chem* 92 (18): 12379–12386.

12 Takeo, E., Sugiura, Y., Ohnishi, Y., Kishima, H., Fukusaki, E., and Shimma, S. (2021). Mass spectrometric enzyme histochemistry for choline acetyltransferase reveals de novo acetylcholine synthesis in rodent brain and spinal cord. *ACS Chem Neurosci* 12 (12): 2079–2087.

13 Takeo, E., Sugiura, Y., Uemura, T., Nishimoto, K., Yasuda, M., Sugiyama, E. et al. (2019). Tandem mass spectrometry imaging reveals distinct accumulation patterns of steroid structural isomers in human adrenal glands. *Anal Chem* 91 (14): 8918–8925.

14 Young, R.S.E., Claes, B.S.R., Bowman, A.P., Williams, E.D., Shepherd, B., Perren, A. et al. (2021). Isomer-resolved imaging of prostate cancer tissues reveals specific lipid unsaturation profiles associated with lymphocytes and abnormal prostate epithelia. *Front Endocrinol (Lausanne)* 12: 689600.

15 Harada, T., Yuba-Kubo, A., Sugiura, Y., Zaima, N., Hayasaka, T., Goto-Inoue, N. et al. (2009). Visualization of volatile substances in different organelles with an atmospheric-pressure mass microscope. *Anal Chem* 81 (21): 9153–9157.

16 Shimma, S., Takashima, Y., Hashimoto, J., Yonemori, K., Tamura, K., and Hamada, A. (2013). Alternative two-step matrix application method for imaging mass spectrometry to avoid tissue shrinkage and improve ionization efficiency. *J Mass Spectrom* 48 (12): 1285–1290.

17 Shimma, S., Kumada, H.O., Taniguchi, H., Konno, A., Yao, I., Furuta, K. et al. (2016). Microscopic visualization of testosterone in mouse testis by use of imaging mass spectrometry. *Anal Bioanal Chem* 408 (27): 7607–7615.

18 Nishimoto, K., Nakagawa, K., Li, D., Kosaka, T., Oya, M., Mikami, S. et al. (2010). Adrenocortical zonation in humans under normal and pathological conditions. *J Clin Endocrinol Metab* 95 (5): 2296–2305.

19 Clift, C.L., McLaughlin, S., Munoz, M., Suuronen, E.J., Rotstein, B.H., Mehta, A.S. et al. (2021). Evaluation of therapeutic collagen-based biomaterials in the infarcted mouse heart by extracellular matrix targeted MALDI imaging mass spectrometry. *J Am Soc Mass Spectrom* 32 (12): 2746–2754.

20 Kip, A.M., Valverde, J.M., Altelaar, M., Heeren, R.M.A., Hundscheid, I.H.R., Dejong, C.H.C. et al. (2022). Combined quantitative (phospho)proteomics and mass spectrometry imaging reveal temporal and spatial protein changes in human intestinal ischemia-reperfusion. *J Proteome Res* 21 (1): 49–66.

21 Lin, M., Eberlin, L.S., and Seeley, E.H. (2022). Reduced hemoglobin signal and improved detection of endogenous proteins in blood-rich tissues for MALDI mass spectrometry imaging. *J Am Soc Mass Spectrom* 33 (2): 296–303.

22 Paulini, F., Araujo, M.S., Silva, L.P., and Lucci, C.M. (2022). Initial steps on mapping differentially expressed proteins in bovine preantral follicles and ovarian tissue: An approach using single-follicle MALDI-MS and mass spectrometry imaging (MSI) analysis. *Reprod Domest Anim* 57 (1): 19–32.

23 Sisley, E.K., Hale, O.J., Styles, I.B., and Cooper, H.J. (2022). Native ambient mass spectrometry imaging of ligand-bound and metal-bound proteins in rat brain. *J Am Chem Soc* 144 (5): 2120–2128.

24 Hoiem, T.S., Andersen, M.K., Martin-Lorenzo, M., Longuespee, R., Claes, B.S.R., Nordborg, A. et al. (2022). An optimized MALDI MSI protocol for spatial detection of tryptic peptides in fresh frozen prostate tissue. *Proteomics* 22 (10): e2100223 (accessed May 2022).

25 Shimma, S., Sugiura, Y., Hayasaka, T., Zaima, N., Matsumoto, M., and Setou, M. (2008). Mass imaging and identification of biomolecules with MALDI-QIT-TOF-based system. *Anal Chem* 80 (3): 878–885.

26 Cordeiro, Y.G., Mulder, L.M., van Zeijl, R.J.M., Paskoski, L.B., van Veelen, P., de Ru, A. et al. (2021). Proteomic analysis identifies FNDC1, A1BG, and antigen processing proteins associated with tumor heterogeneity and malignancy in a canine model of breast cancer. *Cancers (Basel)* 13 (23).

27 Karayel-Basar, M., Uras, I., Kiris, I., Sahin, B., Akgun, E., and Baykal, A.T. (2022). Spatial proteomic alterations detected via MALDI-MS imaging implicate neuronal loss in a Huntington's disease mouse (YAC128) brain. *Mol Omics* 18 (4): 336–347. doi: 10.1039/d1mo00440a.

28 Gomori, G. (1939). Microtechnical demonstration of phosphatase in tissue sections. *Proc Soc Exp Biol Med* 42 (1): 23–26.

6

Elucidation of Protein Function Using Single-molecule Monitoring by Quantum Dots

Maitrayee U. Trivedi[1], *Deepika Sharma*[1], *Alisha Lalhall*[2,3], *Rohit K. Sharma*[1], *and Nishima Wangoo*[2,*]

[1] *Department of Chemistry & Centre for Advanced Studies in Chemistry, Panjab University, Chandigarh, India*
[2] *Department of Applied Sciences, University Institute of Engineering and Technology (U.I.E.T), Panjab University, Chandigarh, India*
[3] *Centre for Nanoscience and Nanotechnology, Panjab University, Chandigarh, India*
[*] *Corresponding author*

6.1 Introduction

Cells, the fundamental unit of life, are composed of a plethora of molecular machines and intricate networks of biochemical reactions [1]. Biomolecules in cells are structurally and dynamically heterogeneous, with transient and active changes in their locations, conformations, and kinetics in different cellular environments and functional states. The behavior of molecules in reactions is critical to biological processes [2, 3]. The ability to detect a single molecule or a small number of molecules selectively in living cells is a powerful tool for addressing many fundamental questions in cell biology. Understanding of how biomolecules behave in their native cellular environment would be greatly enhanced by observing the dynamics of biomolecules in living cells in real time, such as molecule translocation and interactions with the surrounding microenvironment [4]. Many important cellular processes are facilitated by a low number of molecules and thus can be categorized as single-molecule events. Monitoring and tracking the behavior of a single molecule in living cells, e.g. a single protein, is thus a viable strategy for delving into the details of cellular events, which aids in the development of new therapeutic strategies, which in turn further aid in the development of additional therapeutic strategies.

Analytical Techniques for the Elucidation of Protein Function, First Edition. Edited by Isao Suetake, Rohit K. Sharma, and Hironobu Hojo.

The most commonly used technique for single-molecule detection is based on fluorescence. This is because of its high sensitivity and spatial and temporal resolution. In this technique, the molecule of interest is labeled with a fluorophore (fluorescent material) and observed under a microscope [5]. The fluorophores used should be highly photostable and fluorescent, nontoxic in nature, and should not exhibit blinking for super-resolution microscopy. A variety of fluorophores are being used for such applications such as fluorescent proteins, organic dyes, and quantum dots (QDs) and labeled using techniques such as chemical labeling and antibody labeling, protein tags, or unnatural amino acids. Many prevailing bioimaging techniques use green fluorescent protein (GFP) that allow real-time analysis of protein movements in cells. Endosome, endocytosis tracking, and various infectious processes, in particular, have been revealed by single-molecule measurement techniques that use scanning near-field optical microscopy. However, GFP has poor fluorescence and a short fluorescent lifetime (a few seconds to tens of seconds). Therefore, the current decade has witnessed the use of QDs for single-molecule monitoring. QDs have a ten to hundred times greater fluorescence in comparison to dyes such as GFP and are a hundred to thousand times more stable toward photobleaching.

Moreover, QDs are widely used because they have numerous advantages over traditional fluorophores [6]. These fluorescent nanoparticles have broad absorption and narrow emission spectra, a high quantum yield (QY), high fluorescence with a long lifetime, high photochemical stability, and as already explained, are resistant to photobleaching. Conventional fluorophores have drawbacks such as sensitivity to chemical changes within medium, fixed emission spectra, and limited Stokes shift [7]. Additionally, the fact that QDs can emit light in the infrared and near-infrared regions, where tissue absorption is minimal, gives them a significant advantage over conventional fluorophores.

6.1.1 Introduction to Quantum Dots

In the early 1980s, Alexei Ekimov discovered QDs in a glass matrix [8]. QDs basically are fluorescent, inorganic, zero-dimensional semiconductor nanocrystals with sizes ranging from 1 to 10 nm. They have the ability to absorb high-energy light of one wavelength and efficiently convert it to low-energy light of other specific wavelengths [9]. QDs exhibit molecule-like behavior at such small scales by interacting with light via electronic transition dipoles and are thus sometimes referred to as "artificial atoms" because they have discrete electronic energy levels, just like single atoms. QDs have different optical and electrical properties from those of bulk materials due to their smaller size, high surface area, quantum confinement, and dielectric confinement effect. High surface area enhances the efficiency of energy transfer properties, and as a result, QDs have high quantum yield in comparison to other fluorophores. QDs emit light for a longer duration of time after being exposed to a UV energy source, making them a better alternative than other fluorophores. QDs are distinct in that their semiconductor energy levels can be designed simply by changing their size, shape, and charge potential. Because of these energy levels, different-sized QDs have a distinct color. Smaller dots emit higher energy light that is bluer in color, whereas larger dots emit lower energy red light. Typically, 2 nm QDs emit blue light, 3 nm QDs emit green light, and 6 nm QDs emit red light. The quantum confinement effects occur when the size of the nanoparticle is smaller than exciton Bohr radius. In a semiconductor, an exciton Bohr radius is the distance between electrons and holes. When the size of a nanoparticle is smaller than the exciton Bohr radius, quantum confinement effects occur. The physical dimensions of QDs are smaller than the exciton Bohr radius. The electron's energy levels become discrete as a result of quantum confinement, with a finite separation between them. But even so, there are some energy levels that electrons cannot occupy, which is referred to as the band gap. The majority of electrons occupy energy levels below this band gap known as the valence band; in fact, almost all energy levels in the valence band are occupied. But if an external stimulus is applied, an electron may shift from the valence band to the conduction band (energy levels above the band gap) [10]. When the incident light strikes the QDs, they absorb a photon with higher energy than the band gap of the constituting semiconductor. When the electron returns to a lower energy level, it emits a narrow, symmetric energy band. The radiation emitted can be measured to determine the band gap size of a semiconductor. The wavelength of photon emission is affected not only by the material used to make the QDs, but also by their size. The greater the size of the QDs, the smaller the band gap and the more continuous its electronic structure, which emits photons with long wavelengths (red shift), similar to bulk materials, whereas smaller QDs have a larger energy band gap and thus emit light with a shorter wavelength (blue shift) as seen in Figure 6.1a. The size dependence demonstrates that the properties can be easily tuned by varying the size of the QDs. Quantum confinement in QDs results in unique optoelectronic properties such as high QY and narrow emission bands that can be tuned by increasing or decreasing the size of the QDs [11].

6.1.2 Types of Quantum Dots

QDs are categorized according to their chemical composition and are made up of elements from periodic table groups III–V, II–VI, or IV–VI (Figure 6.1b). QDs are typically classified into three types: (i) core-type QDs, (ii) core/shell-type QDs, and (iii) alloyed-type QDs [10].

6.1.2.1 Core Type QDs

Core-type QDs are composed of a single material with uniform internal compositions, usually a chalcogenide such as sulfides, selenides, or tellurides of metals

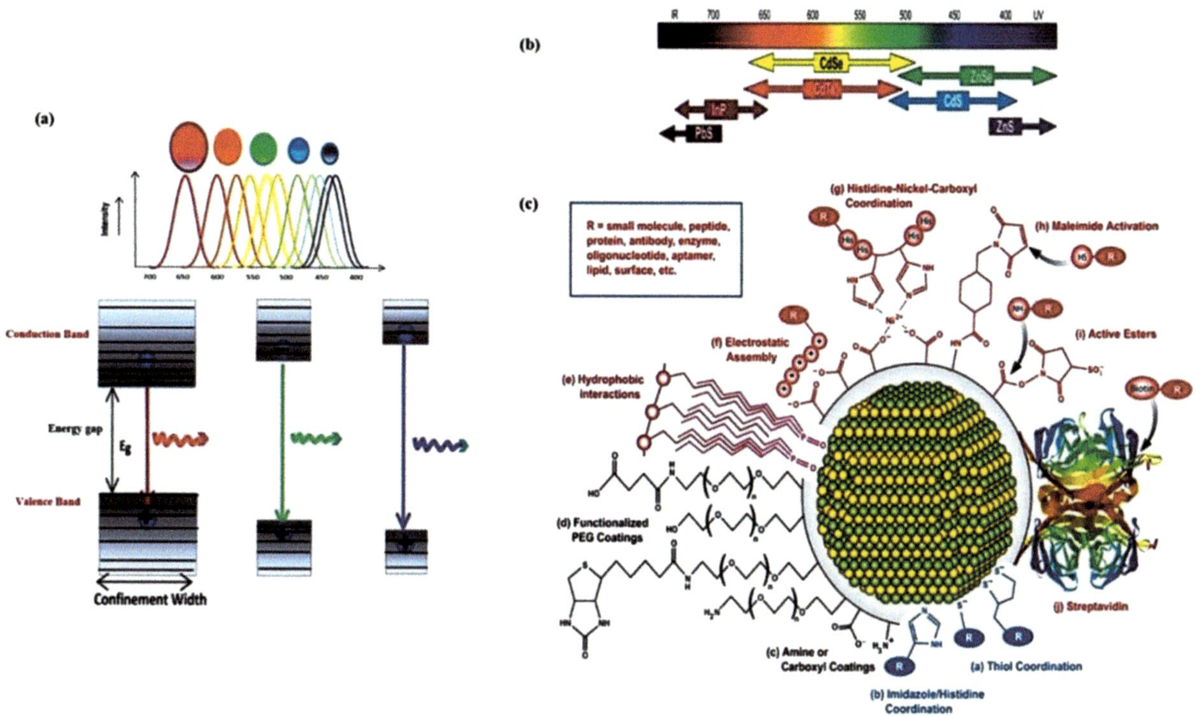

Figure 6.1 (a) Varying band gaps with varying QDs size. With permission of Elsevier. (b) Different emission spectra of different types of QDs. [13] / Tabriz University of Medical Sciences. (c) Various bioconjugation techniques. [14] / With Permission of Elsevier. Reprinted from Analytica Chimica Acta, Vol. 673, W. Russ Algar, Anthony J. Tavares, Ulrich J. Krull, "Beyond labels: A review of the application of quantum dots as integrated components of assays, bioprobes, and biosensors utilizing optical transduction," 1–25, Copyright (2010), with permission from Elsevier.

such as cadmium, lead, or zinc. By simply varying the crystallite size, the photo- and electroluminescence properties of core-type nanocrystals can be tuned. CdSe, CdTe, PbS, and ZnS are a few examples.

6.1.2.2 Core/shell-type QDs

A luminescent semiconductor core is capped by a thin shell of higher band gap semiconducting material in core/shell-type, or core-shell, QDs. The core and shell are made of semiconductors of type II–VI, IV–VI, and III–V, such as CdS/ZnS, CdSe/ZnS, CdSe/CdS, CdTe/CdS, and InAs/CdSe. The QDs surface is a dynamic chemically heterogeneous interface between crystalline solid atoms and ligands coordinated to the surface atoms. Due to the overall high surface area–to-volume ratio and unsaturated or dangling bonds, which exist on the surface, these under-coordinated atoms make QDs more active. The surface is exposed to solvents and other species in solution, as well as matrices in its surroundings, which can significantly affect the properties of QDs [15]. Surface atoms' lower coordination compared to bulk atoms may result in localized electronic states or highly reactive sites susceptible to a variety of chemical and redox processes. Shallow or deep mid gap states as surface traps capable of providing pathways for nonradiative exciton recombination are highly probable in these conditions, which is detrimental to various QDs-based applications. Shallow traps are states that are close to the band edges (either conduction band or valence band), whereas deep traps are states that are close to the middle of the band gap. So organic ligands can be used as capping agents, to passivate surface traps, for stabilization, and for better dispersibility in the liquid medium. However, the primary obstacle in using organic ligands is the complexity in simultaneously passivating anionic and cationic surface traps, as well as the steric hindrance between bulky organic ligands, which results in incomplete surface coverage and un-passivated dangling orbitals [16]. To eliminate these surface traps, a shell of a wider band gap material is grown around the core, forming core/shell QDs and thus relocating surface defects to the inorganic shell's outer region. The shell growth protects against photo-oxidative and chemical degradation, as well as environmental changes, and provides an additional approach for functionalization of the QDs. In addition, coating also results in higher quantum yield and greater stability than core QDs. Therefore, core/shell-type QDs are the most stable type [17].

Core/shell QDs are classified into three broad groups based on the band gap and energy levels of their constituents: type I, reverse type I, type II, and reverse type II [10]:

Type I: In this type, either the core's conduction or valance band aligns within the band gap of the shell, enabling electrons and holes to be localized in the core. In CdSe/CdS type-I, for example, the CdSe core has a band gap of 1.74 eV and the CdS shell has a band gap of 2.42 eV. As a result, both holes and electrons are constrained to the CdSe core. The average absorption range is 400–500 nm with average emission range 430–600 nm. Other examples of type-I QDs include CdSe/ZnS and InAs/CdSe.

Inverse type I: In inverse type-I, the band gap of the core is wider than the band gap of the shell, and so both the conduction and valence bands of the shell are localized within the band gap of the core. As a result, the holes and electrons are confined in the shell. The average absorption range is 400–500 nm with average emission range 400–700 nm. Examples include CdS/HgS, CdS/CdSe, and ZnSe/CdSe.

Type II: In type II, the core's valence and conduction band edges are lower or higher than those of the shell, and both the hole and the electron are confined to the core. Examples of type-II are CdS/HgS, CdS/CdSe, and ZnSe/CdSe. Unlike type-I QDs, where the band gap is determined solely by the core, the band gap of type-II QDs is determined by the energy level difference between the core and the shell. The average absorption range is 600–800 nm with average emission range 700–1000 nm.

Inverse type II: In this type, the core's conduction band edge is within the shell's band gap, or the shell's valence band edge is within the core's band. They have high quantum yield and fair stability. The average absorption range is 300–1600 nm with average emission range 700–1000 nm. Examples include InP/CdS and PbS/CdS.

6.1.2.3 Alloyed-type QDs

Alloyed-type QDs are created by alloying two semiconductors with different band gap energies, resulting in properties that differ not only from their bulk counterparts but also from their parent semiconductors. These QDs with both homogeneous and gradient internal structures allow optical and electronic properties to be tuned by simply altering the composition and internal structure without changing the crystallite size. For example, Zn (Te1−xSex) and CdSxSe1-x/ZnS.

6.2 Synthesis Methods

QDs have been synthesized using a variety of methods as seen in Table 6.1. In general, the techniques for synthesis of QDs are classified as either top-down or bottom-up [18]. Top-down approach: Top down is a physical approach in which a bulk semiconductor is broken or thinned to form QDs. The top-down approach includes etching, reactive ion etching, and focused ion beam etching. The bottom-up, or self-assembly, approach uses chemical or physical forces operating at the nanoscale level to assemble basic units into larger structures. This approach can be broadly divided into wet-chemical and vapor-phase methods, which are described below.

Table 6.1 Various techniques for the synthesis of QDs.

Top-down approach	
Etching	• A reactive gas species is injected into an etching chamber, and a radio frequency voltage is applied to generate plasma, which breaks down the gas molecules into more reactive fragments. • These high kinetic energy species collide with the surface, forming a volatile reaction product that etches a patterned sample.
Reactive ion etching (RIE)	• Plasma ions with high energies attack and react with the wafer surface, removing the material that has been deposited on the wafer. • Plasma is generated under low pressure (vacuum) by an electromagnetic field.
Focused ion beam (FIB)	• Highly focused beams from a molten metal source (e.g. Ga, Au/Si, Au/Si/Be, or Pd/As/B) can be used directly to sputter the surface of the semiconductor substrate. • Is a slow, low-throughput method that involves expensive equipment and results in residual surface damage.
Bottom-up approach (a) Wet chemical methods	
Sol gel	• A sol (nanoparticles dispersed in a solvent by Brownian motion) is prepared in an acidic or basic medium using a metal precursor (alkoxides, acetates, or nitrates). • Metal precursor hydrolyzes in the medium and condenses to form a sol, followed by polymerization to form a network (gel). • Disadvantages: broad size distribution and high concentration of defects.
Microemulsion	• Types: (a) normal microemulsions, i.e. oil-in-water; (b) reverse microemulsions, i.e. water-in-oil. • Reverse micelle is widely used, where two immiscible liquids (polar water and nonpolar long-chain alkane) are mixed and stirred to form emulsion. • Advantages: easy control of the size by changing the molar ratio of water to surfactant, a narrow distribution of size, and ease of dispersion. • Disadvantages: low yield and incorporation of impurities and defects.
Hot solution decomposition	• High temperature (~300 °C) pyrolysis of organometallic compound. • Following degassing, a mixture of the precursor and a coordinating solvent is prepared under vigorous stirring at ~300 °C, resulting in homogeneous nucleation to form QDs and subsequent growth via Ostwald ripening. • Advantages: sufficient thermal energy to anneal defects and monodispersity. • Disadvantages: higher costs, toxicity of the organometallic precursors, and poor dispersion in water.

(Continued)

Table 6.1 (Continued)

Microwave	• Microwaves are passed through a mixture of precursors, providing energy to dissociate the precursor and water molecules in water to grow QDs. • Through adiabatic compression within the gas inside the collapsing bubble, such acoustic cavitation creates a localized hotspot, enabling the reactions that form QDs.
Hydrothermal	• Crystallization of inorganic salts from aqueous solution, controlled by pressure and temperature. • As the temperature and/or pressure are reduced, the solubility of inorganic compounds decreases, resulting in crystalline precipitates. • Different shapes and sizes of QDs can be achieved by varying reactants, pressure, temperature, and reaction and aging time.
Bottom-up approach (b) Vapor phase methods	
Physical vapor deposition (PVD)	• Condensation of a solid from vapors produced by thermal evaporation or by sputtering. • Evaporation techniques: electron beam heating, resistive or Joule heating, arc-discharge, and pulsed laser ablation.
Chemical vapor deposition (CVD)	• Precursors are introduced into a chamber at a specific pressure and temperature, diffuse to the heated substrate, react to form a film, and then gas-phase by products desorb from the substrate and are removed from the chamber.
Molecular beam epitaxy (MBE)	• Overlayers and elemental growth, compound, or alloy semiconductor nanostructured materials are deposited on a heated substrate under ultra-high vacuum condition. • Evaporation from an apertured source to form a beam of atoms or molecules. • Beams formed from solids (e.g. elemental Ga and As) or a combination of solid plus gases (e.g. AsH3, PH3, or metal-organics such as tri-methyl gallium or tri-ethyl gallium).

6.2.1 Wet-chemical Methods

Wet-chemical methods are conventional precipitation methods in which parameters for a single solution or mixture of solutions are controlled. Nanoparticle nucleation and limited growth are both entailed in the precipitation process. QDs of the desired size, shape, and composition can be achieved by varying parameters such as temperature, thickness, stabilizers or micelle formation, precursor concentrations, anionic to cationic species ratios, and solvent.

6.2.2 Vapor-phase Methods

The layered material grows as a uniform, frequently epitaxial layer, beginning as a smooth layer and sometimes followed by nucleation and growth. Even though the vapor-phase method is effective in fabricating self-assembled QDs arrays without a template, fluctuation in size frequently results in inhomogeneous optoelectronic properties.

6.3 Bioconjugation

QDs are one of the most efficacious nanoprobes for any bio-application, including chemical, biomedical, and therapeutic labeling and imaging, and cell targeting. However, for these types of applications, the particles must meet certain criteria, such as (i) being stable in aqueous solutions over a wide pH and ionic strength range; (ii) retaining their optical properties; (iii) have functional groups on their surface that are available for conjugation; (iv) Solubilization of QDs enables further conjugation to biomolecules such as proteins, immunoglobulins (Igs), aptamers, and oligonucleotides, resulting in specific binding, leading to a significant improvement in QDs' specificity. Bioconjugation tends to result in multifunctional QDs that combine the optical/electrochemical properties of QDs with a biomolecule's biological function. QDs are bioconjugated with antibodies, proteins, peptides, aptamers, nucleic acids, small molecules, and liposomes via covalent or noncovalent conjugates, which are ubiquitously seen in direct and indirect labeling of extracellular proteins and cellular components (Figure 6.1c).

There are three types of conjugation schemes for attaching proteins to QDs: (i) condensation of EDC, 1-ethyl-3- (3-dimethylaminopropyl) carbodiimide, to react carboxy groups on the QDs surface to amines; (ii) direct binding to the QDs surface using thiolated peptides (dative bond between QDs surface sulfur atoms and cysteine residues) or polyhistidine (polyHis) residues (metal-affinity coordination of His residues to QDs surface Zn atoms); and (iii) adsorption or noncovalent self-assembly using engineered proteins (using EDC) [19]. Nonspecific QDs-cell membrane binding takes place as a result of hydrophobic and electrostatic interactions between QDs capping molecules and biomolecules in the cell membrane. Furthermore, the surface of core or core/shell QDs must contain a reactive functional group or molecule such as streptavidin, biotin, primary amine, thiol, maleimide, succinimide, or carboxylic acid to facilitate biolabeling. In bio-imaging applications, for example, QDs can serve as imaging tags, while the attached Igs may act as a unique targeting agent through specific antigen binding action [20].

Wu et al. demonstrated the intracellular and intranuclear delivery of bioconjugated multicolor CdSe/ZnS QDs in human epithelial cells or 3T3 cells by incubating fixed cells with primary antibody, biotinylated secondary antibody, and QD-streptavidin conjugate. Bioconjugated QDs were developed by coating CdSe/ZnS QDs with poly-(acrylic acid), cross-linking PAA with lysine or PEG-lysine, and conjugating lysine to streptavidin [21]. Xiaohu Gao et al. developed bioconjugated QD probes for in vivo targeting and imaging of human prostate cancer cells in mice. An amphiphilic triblock copolymer for in vivo protection, targeting-ligands for tumor antigen recognition, and multiple PEG molecules were used [22].

6.4 Analytical Methods for Single-molecule Monitoring by Quantum Dots

The measuring and monitoring techniques used for QD-based single-molecule monitoring are described in the following section of the chapter. Several optical imaging techniques have been developed so far mainly including two types: (i) wide-field imaging microscopic techniques, which are categorized into epifluorescence microscopy, total internal reflection fluorescence microscopy (TIRFM), and quasi-TIRFM, and (ii) the confocal microscopic technique, which includes laser scanning confocal microscopy (CLSM), spinning-disk confocal microscopy (SD-CLSM), and single-point edge excitation sub-diffraction microscopy (SPEED). These microscopic techniques have been developed in recent decades and have demonstrated their ability to capture single-molecule dynamics in living cells. In this section we discuss future developments and briefly summarize recent progress in using these imaging techniques for monitoring protein function for a better understanding of important biological processes. The microscopic techniques are described in the order in which they were developed.

6.4.1 Epifluorescence Microscopy

Epifluorescence microscopy, also known as wide-field microscopy (WFM), is one of the most popular fluorescence microscopy techniques in the field of life sciences. Epifluorescence microscopy generally uses a laser for fluorescence excitation and a highly sensitive CCD for detecting the signal of a single molecule. The construction of an epifluorescence microscope is given in Figure 6.2a. In brief, an arc-discharge lamp or any other light source is used to generate multispectral light. The generated light is passed through a wavelength selective bandpass filter known as excitation filter. Only a specific wavelength of light, normally in the UV or blue–green regions of the visible light spectrum, passes through the excitation

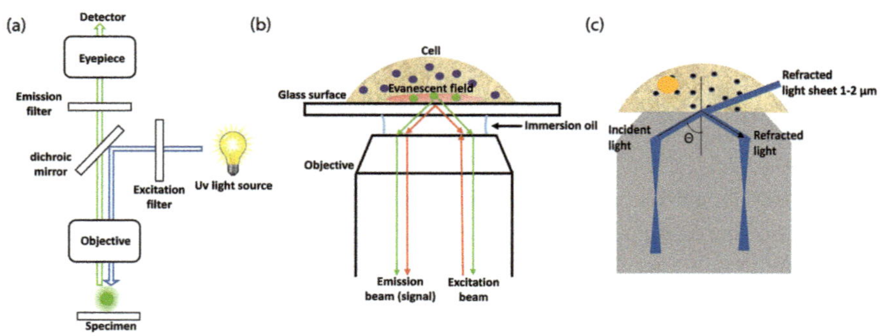

Figure 6.2 (a) The construction of epifluorescence microscopy, (b) TIRFM, and (c) quasi-TIRFM.

filter. This light is then reflected from a dichromatic mirror or beam splitter through the microscope objective exposing the sample to intense light. If the specimen shows fluorescence, the emitted light collected by the objective passes through the dichromatic mirror and then is filtered by another bandpass filter, i.e. the emission filter, where the unwanted excitation wavelengths are blocked. Finally, a detector, such as a CCD camera, collects the emitted light.

In epifluorescence microscopy, you can observe primary neurons, cell morphology, cellular/subcellular compartments, and cellular markers of disease, such as cancer compared to normal cells or phenotypes such as stem cells. The limitation of epifluorescence is because of the interference arising from the out-of-focus fluorophores causing blurring of images. In order to enhance the signal-to-noise ratio and detect single molecules, strong fluorescent emitters such as QDs or dyes are applied [23]. In order to decrease the cell background signal, very small QDs near-infrared fluorescence emission was made that would be helpful in in vivo imaging [24]. Several advancements in filters, cameras, and light sources have allowed scientists to see the microscopic functioning of organisms, cells, and tissues. Microscopic techniques such as TIRFM and confocal microscopy are utilized for reducing the excitation/emission volume hence resulting in decreased background signal for single-molecule monitoring.

6.4.2 Total Internal Reflection Fluorescence Microscope

Total internal reflection fluorescence microscopes (TIRFM) (Figure 6.2b) are used to observe a thin section of a specimen, typically less than 200 nm. In this technique, fluorescence cells are excited in a thin optical specimen section mounted on a glass slide. This method works on the principle of total internal reflection. When light waves reach a boundary between two media (glass and water), instead of refracting into the second medium, they are reflected back into the first medium, known as total internal reflection. In TIRFM, total reflectance occurs when excited light passes through a medium such as a transparent solid cover glass with high refractive index into water with low refractive index. During this process, only an electromagnetic radiation known as the evanescent wave passes through the low refractive index medium [25]. The frequency of the evanescent wave decays exponentially along with the distance from the solid's surface, and only fluorescent molecules around a few 100 nm of the solid are proficiently excited. From these 2D images of the fluorescence a 2D image can be obtained, although there are ways that can determine 3D information on the location of vesicles or structures in cells. TIRFM can distinctively envision interface regions of cells cultured on a cover glass, such as the basal plasma membrane. Since many cellular events, such as endocytosis, ligand receptor binding, and viral infection, begin at the plasma membrane, TIRFM has turned out to be the most important tool for studying single molecules in living cells [26, 27].

6.4.3 Confocal Microscopy

Even though TIRFM is a strong technique for molecular imaging closer to the cell surface, molecules and structures inside cells are outside its reach. For molecular imaging of processes occurring inside the cytoplasm, confocal microscopy is a decent option. It includes utilizing a pinhole in detecting a path for removing out-of-focus signals arising from fluorescence in specimens. There are two types of confocal microscopy techniques responsible for single-molecule monitoring available, i.e. laser scanning (CLSM) (Figure 6.3a) and spinning-disk confocal microscopy (SD-CLSM) (Figure 6.3b). In CLSM fluorescence is generated from a tiny (approximately 1 μm^2) excited area of the sample. An image is recorded by scanning the laser beam through the sample in a faster scanning mode, and photons are recorded pixel by pixel. CLSM produces sharp optical slices of samples being imaged. However, if each slice is partly overlapped along the z dimension and the image stacks are recorded, it is used for generating three-dimensional images of the sample. This has been very famous in research of viruses for localizing individual viruses in various compartments of static and live cells [28]. It could also be employed for molecular number counting relevant to single-molecule imaging.

CLSM has a limitation since point-to-point signal gathering takes a significant amount of time to acquire clear images. The image frame rate could be enhanced by SD-CLSM [29]. SD-CLSM utilizes two coupled disks each having numerous arrayed pinholes along with microlenses. A single pair of pinhole microlens behaves as one reciprocal-geometry confocal system. On spinning of these disks, an array of collimated focus laser beams scan the sample leading to imaging of the complete specimen and enhancing the image collecting speed. The photons emitted from the points scanned are simultaneously detected by using more than one electron-multiplying charge-coupled devices (EMCCD) camera. SD-CLSM is a well-proven technique for monitoring viruses in real time inside cells [30].

Even though the signal-to-noise ratio is significantly enhanced in comparison to epifluorescence microscopy, the time resolution in confocal microscopy is not up to the mark. Similar to the epifluorescence imaging technique in confocal microscopy if an appropriate fluorescent probe is not available, the single-molecule monitoring is disturbed. For the single-molecule monitoring to be feasible, the brightness and stability of fluorescent probes are a must.

6.4.4 pseudo-TIRFM

Since confocal microscopy is restricted due to its comparatively low time resolution and TIRFM is limited due to the processes taking place on the cell membrane, this restricts their use in individual viral or molecule imaging within three-dimensional samples such as cells and cellular compartments. To overcome these limitations, there is a need to balance the signal-to-noise noise ratio occurring from TIRFM and

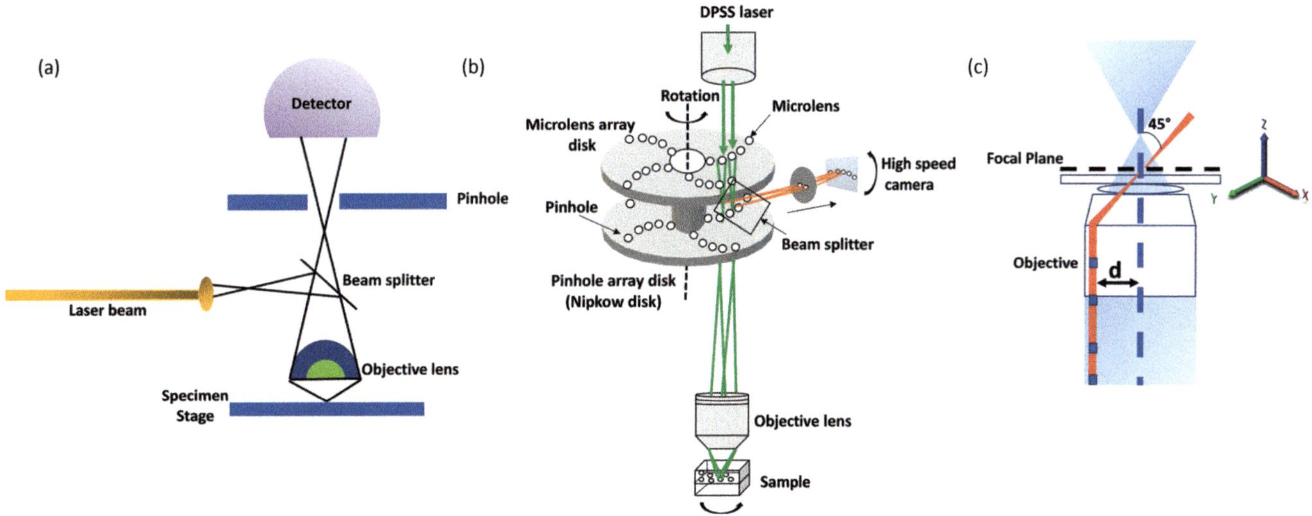

Figure 6.3 (a) Laser scanning confocal microscopy (CLSM), (b) spinning-disk confocal microscopy (SD-CLSM), and (c) SPEED.

incorporate three-dimensional imaging [31]. Keeping these points in mind two new techniques were developed: pseudo-TIRFM (quasi-TIRFM) (Figure 6.2c) and single-point edge excitation subdiffraction microscopy (SPEED). This has enabled to trap intracellular processes at videography rate. When the incident angle was set a little bit lesser than the critical angle, i.e. between epi-illumination and TIRFM, quasi-TIRFM refracts the excitation light within the sample in the form of a 1–2 μm sheet. As the molecules away from the sheet are rarely excited, it allows the decrease in background signal arising from the cytoplasm [32]. On utilizing this methodology for illumination, QD-nerve growth factors located inside neuron bodies and axons over hundreds of nanometers far from the cell were tracked in real time by Cui et al. [33]. They were able to determine that the single endosome containing QD-nerve growth factor had "stop and go" along with single direction retrograde movement having an average speed ranging from 0.2 to 3 $\mu m\ s^{-1}$. Since the maximum of endosomes had presence of a single QD-nerve growth factor, their research indicated probability for the presence of a single QD-nerve growth factor dimer being sufficient for sustaining signaling at the time of retrograds axonal transport to the cell body. In a separate study, pseudo-TIRFM and microfluidic techniques were clubbed and the researchers found that on incubating QD-nerve growth factor molecule with neuron axon, the image showing active transport image of one QD-nerve growth factor from a faraway axon compartment to the neuron body can be captured [32]. Both of these works have imparted light on analysis of axonal transport processes.

6.4.5 Single-point Edge Excitation Subdiffraction Microscopy

SPEED microscopy (Figure 6.3c) uses single-point illumination and detection, similar to CLSM but with some changes [34]: (i) Diffraction-limited spots are observed using CCD camera installed in SPEED, (ii) a beam of laser shifted off the center from the objective is focused on inclined illumination volume with a diffraction angle of 45° within the focal plane, and a pinhole is not used in the emission path, (iii) since it does not use a pinhole like the CLSM technique in the emission path, this allows the detector and CCD camera to get a higher number of signals while maintaining a satisfactory background fluorescence. Such changes allow the CCD camera to identify the signal from lower background fluorescence [34].

6.5 Applications

6.5.1 Application of Single-molecule Monitoring Using QD for Enlightening Nanoscale Neuroscience

The regular functioning of the central nervous system (CNS) is efficiently controlled by neurotransmitters (endogenous chemicals secreted by neurons) that modulate signal processes between nerve cells and effector organs [35]. The stimulation and inhibition of neuronal transmission in the CNS involves the neurotransmitter release at presynaptic neurons followed by postsynaptic receptor activation [36]. In general, there are two types of synapses (the correlating points between two neurons where information is passed): excitatory and inhibitory synapses. An excitatory synapse causes depolarization of a postsynaptic neuron or cell as a result of decrease polarity of cells owing to influx of sodium cations [37]. The inhibitory synapses lead to hyperpolarization of cells by uptake of anions or efflux of sodium ions. Glycine and GABA (gamma-aminobutyric acid) are examples of inhibitory neurotransmitters [38]. The abnormal functioning of neurotransmitters can disrupt the stability of cells and have adverse effects. Therefore, the issue of lateral mobility of these neurotransmitters has gained attention in order to understand the development and plasticity of synapses.

6.5.2 Investigation of Diffusion Dynamics of Neuroreceptors in Cultured Neurons

Neuroreceptors are membrane proteins that are activated by neurotransmitters. The lateral diffusion of neuroreceptors has been investigated using QD single-particle tracking (SPT). Such system can be investigated without the need of intracellular delivery across the impermeable plasma membrane. Dahan et al. were first to utilize QDs as fluorescent probe to track individual glycine receptor (GlyR) and its lateral diffusion in neuronal membrane of living cells that last for 20 mins. [39]. GlyR are pentameric ligand-gated ion channels whose key function is to inhibit fast neurotransmission in the adult brainstem and spine. QD-tagged GlyR receptor was used to analyze multiple diffusion domains of GlyR localization in synaptic or perisynaptic domains. Owing to the similarity in size of QD labeled GlyR and the synaptic cleft, QDs labeled to GlyR were easily visualized while entering and existing in the confined environment. Charrier et al. also tracked the lateral diffusion of GlyR using QD by measuring the change in concentration between GlyR and gephyrin at synapses in living cells [40]. The gephyrin, scaffolding protein is known to attach to GlyR clusters at synapses and also maintain dynamic equilibrium between synaptic and extrasynaptic GlyR in the neuronal membrane. SPT revealed about cytoskeleton disruption, which led to augmented GlyR exchanges between the synaptic and extrasynaptic membranes as well as reduced receptor residing time at synapse. QD labeled GlyR was utilized to analyze the kinetic parameters of two GlyR subunit types. The findings were helpful in understanding the role of gephyrin-gephyrin interaction for the stabilization of GlyR [41].

Growth cones (GC), known as tip [42] of axon (nerve fiber), determine the direction of growth and then guide the extension of the axon in that direction. This axonal navigation is dependent on extracellular concentration gradients of several guidance signals, including GABA receptor ($GABA_AR$). $GABA_AR$ plays a crucial role in the normal development of brain circuits, along with monitoring of the excitation-inhibition balance, which is essential for proper functioning of brain development throughout life [42]. A single-molecule assay was developed by combining $GABA_AR$ to QDs to investigate the single $GABA_AR$ in plasma membrane of nerve growth cone in the presence of external directional stimuli. The effect of an external GABA gradient has led to asymmetric redistribution of $GABA_ARs$ across the growth cones (GC) toward the gradient source as a result of transient interaction between receptor and microtubules, which was observed using epifluorescence microscopy. The redistribution of $GABA_ARs$ was assisted by an increase in intracellular calcium concentration [43].

6.5.3 Single-molecule Tracking of Neuroreceptors in Intact Brain Slices (in Vivo)

Neuroreceptors functioning at excitatory and inhibitory synapses have not only been studied in cultured neurons (in vitro potent method) but has also been investigated in complex brain slices (in vivo). Although, cultured neurons are suitable systems to study some neural mechanisms for superficial cells, they clearly behave differently in various aspects from cell networks inside intact brain slices. Varela et al. for the first time employed QDs to track and label dopamine receptors in an acute rat brain slice [44]. Dopamine receptors are G-protein coupled receptors that regulate the dopamine in central and peripheral organs. The major function of the mammalian CNS such as locomotion, novelty detection, and long-duration memory development are majorly controlled by dopaminergic signaling [45]. The abnormal or disfunction of dopaminergic signaling can drastically alter the synaptic function and plasticity and further can lead to life-threatening diseases including Parkinson's disease or schizophrenia. A single QD was detected by spinning-disk microscopy operating at a rate of 10–30 frames per second. Surprisingly, the QD was detected even up to a depth of 50 μm inside a brain slice depicting roughly to 3–4 pyramidal cell layers inside the slice (Figure 6.4). In addition, membrane permeability assay was also carried out on artificial cerebrospinal fluid (ACSF) by adding propidium iodide (PI) when imaging brain slices. The experiment has been performed at 350 μm and 50 μm thick acute brain slice, where 3D reconstruction images depicted that only the first layer was composed of stained nuclei in comparison to deeper layers that contained PI-negative healthy cells at 350 μm thickness. At 50 μm thick slices, the entire volume of the slice was covered with PI-positive dead cells. Thus, 350 μm brain slice was subsequently used at depths above 20 μm within a slice. Thus, such tracking tagged receptors strategy could be helpful in understanding the dopamine surface distribution during pathological dysfunction in neuropsychiatric disorders.

In another report, small and single domain antibodies (V_HH only or sdAb) having high affinity toward green fluorescent protein (GPF) recognition were conjugated to QD to monitor the mobility of $GABA_AR$ in an acute brain slice [46].

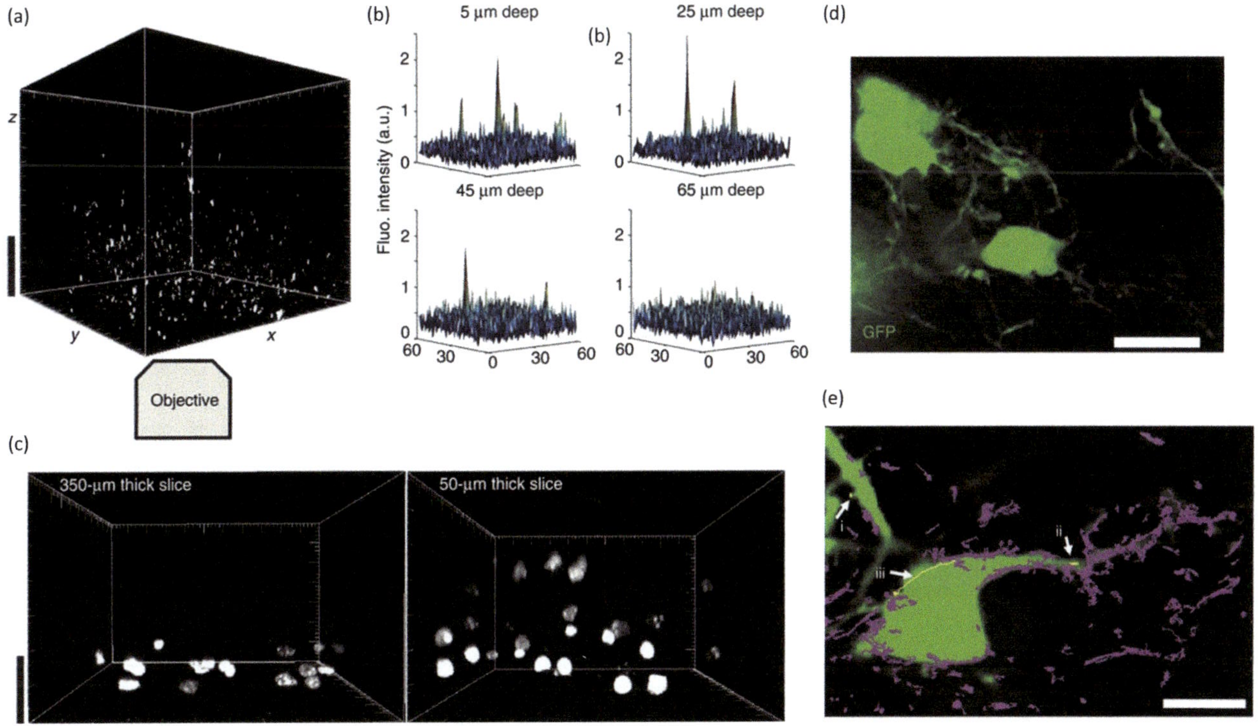

Figure 6.4 (a) Spinning-disk confocal microscopy depicting 3D reconstruction of z-stacking in an acute brain slice of a rat injected with QD (scale bar: 20 μm). (b) The imaged QD at least up to depths of 45 μm can be detected in live tissue. (c) Membrane permeability assays showing a layer of damaged cells in 350 μm slices and in 50 μm thick slices (scale bar: 20 μm). (d) Image from spinning-disk confocal microscope of acute brain slices of electroporated animals when injected with functionalized QD (scale bar: 15 μm). (e) Fast two-dimensional imaging was performed over transfected neurons with coupled QD and subsequently QD were identified and trajectories reconstructed (scale bar: 10 μm). (a–e) [44], Varela et al. (2016) / Springer Nature, CC BY 4.0.

With the aid of widefield microscopy, QD antibody conjugate can be observed inside and outside synapses for a longer time. In addition, this nanoconjugate can be utilized to study the different features of membrane protein dynamics either at established excitatory and inhibitory synapses or during development of axons. Further, the QD-nanobodies were employed to study the mobility of $GABA_AR$ and enhanced the possibility of imaging receptors in intact tissues with high precision. The applicability of the QD-antibody conjugate was also extended for simultaneous regulation of several proteins expressed in similar cells.

6.5.4 QD-tagged Neurotransmitter Transporters

Neurotransmitter transporters belong to the class of membrane transport proteins that mainly carry neurotransmitters across membranes to specific intercellular locations based on an electrochemical gradient across the membrane. They are mostly located in synaptic clefts, where they act to dismiss the effects of the neurotransmitter by removing it from the cleft. Membrane carrier for serotonin (serotonin transporter, or SERT), dopamine (dopamine transporter, or DAT), norepinephrine (norepinephrine transporter, or NET), and GABA (GABA transporters GAT1–3) are some of the examples that belong to neurotransmitter transporter family.

6.5.5 QD Labeled Serotonin Transporter (SERT) to Understand Membrane Dynamics

SERT belongs to the neurotransmitter-sodium symporter (NSS)family, encoded in humans by the SLC6A4 gene. Mutated SERT expression can lead to numerous neuropsychiatric disorders including anxiety, depression, and autism. Single-molecule analysis of presynaptic SERT (5-hydroxytryptamine, 5-HT) was monitored using antagonist QDs on the surface of serotonergic cells [48]. The investigation from initial measurement of single SERT diffusion revealed the existence of SERT in two different pools in the membrane of RN46A cells. The major fraction (90%) displayed highly restricted diffusion, whereas the remaining pool diffuses freely at a much faster rate. Toxin-B, a marker of cholesterol, and GM1 ganglioside rich membrane microdomain was tagged with Alexa488, which depicted high a degree of colocalization of QD-SERT using parallel imaging. Methyl-β-cyclodextrin, a commonly used cholesterol-extracting agent, treatment on SERT-expressing RN46A cells has resulted in dramatic increase in the overall diffusion rate of SERT, along with disruption of GM1 positive membrane microdomains. Also, methyl-β-cyclodextrin treatment indicated complete loss of restricted SERT pool under basal conditions as a consequence of SERT localization to the membrane microdomains. The majority of SERT proteins under basal conditions were located in the putative microdomains approximately 500 nm in diameter, which was analyzed from the

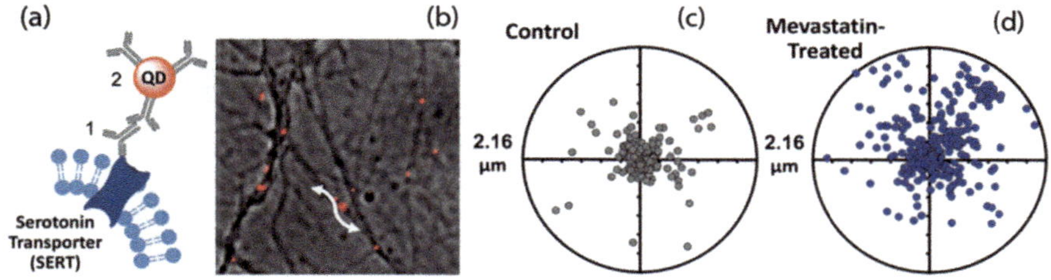

Figure 6.5 (a) QD labeling of SERT by antibodies in a two-step labeling method; firstly, primary antibody is linked, followed by secondary antibody-conjugated QDs; (b) differential interference contrast (DIC) image of QD655 channels to demonstrate binding along neurites. Two-dimensional polar plots of 5 second radial displacements (dr) of single QD-SERT complexes; (c) for control (mean dr = 0.38 ± 0.031 µm); (d) for Mevastatin treated (dr = 0.55 ± 0.028 µm). Radius of a plot is 2.16 µm (12 pixels). [47], Bailey et al. (2018) / with permission from American Chemical Society.

SERT trajectory by applying the 5-second dr. In addition, the restricted SERT pool appeared to be under the fundamental control of the p38 mitogen-activated protein kinase (p38 MAPK) pathway.

In another report, QD-labeled endogenous SERT has been employed to investigate SERT dynamics in primary rat midbrain neurons [47]. Epifluorescence microscopy was employed for the investigation of SERT diffusion dynamics and its correlation with membrane cholesterol and SERT phosphorylation level. It was found that lateral mobility and conformationally sensitive Thr276 in endogenous SERT have been affected by reduced levels of membrane cholesterol as determined from two different methods based on cholesterol manipulation: statins and methyl-β-cyclodextrin treatment (Figure 6.5). Membrane cholesterol plays a significant role in regulating SERT and can lead to changes in membrane fluidity along with specific protein conformation. The increased SERT surface mobility in its native environment was the result of activation of membrane-permeable 8-bromoguanosine 3′,5′-cyclic monophosphate (Br-cGMP)-mediated protein kinase G (PKG). In addition, the phosphorylation of Thr276 in endogenous SERT was enhanced by a factor of 2.5 based on both chronic and acute cholesterol depletion. Thus, QD-SPT is capable of demonstrating the molecular details of pharmacological regulation and catalytic activation of single proteins in living cells occurring at timescales of milliseconds to nanoseconds.

6.5.6 Membrane Trafficking and Imaging of Dopamine Transporter (DAT) Using QDs

Dopamine transporter (DAT, SLC6A3) is a presynaptic, cocaine- and amphetamine-sensitive neurotransmitter transporter. Its key role is to regulate the intensity and length of synaptic dopamine (DA) signals by instantly clearing back DA

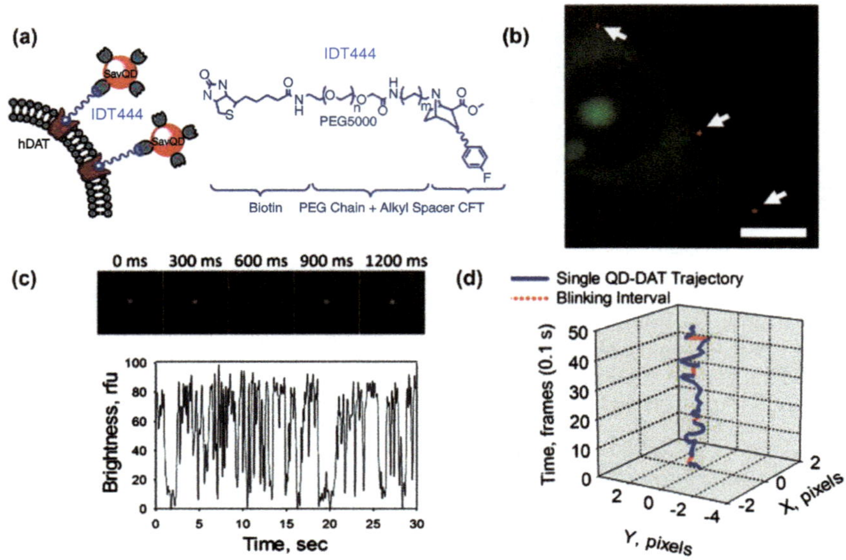

Figure 6.6 Time-lapse imaging of DAT of stably transfected Flp-In 293 cells in the plasma membrane. (a) Schematic representation for QD labeling protocol; firstly, DAT-specific IDT444 cocaine analogue binding with DAT followed by its biotin terminus enabling subsequent DAT recognition with SavQDs. (b) Fluorescence micrograph showing characteristic surface QD labeling of DATs (bar: 5 μm). (c) Representative time-lapse image series of a single SavQD immobilized (spun-cast) on a glass coverslip and a corresponding emission spectrum trace. (d) Single DAT-QD complex trajectory undergoing diffusion in the plasma membrane. [49], Kovtun et al. (2015) / with permission from American Chemical Society.

into presynaptic nerve terminals. The dysfunction of DAT-mediated DA clearance can result in a variety of neuropsychiatric disorders such as addiction, autism, and attention deficit hyperactivity disorder (ADHD). Kovtun et al. have investigated the lateral mobility of single DAT (antagonist)-QD conjugate in living cells to monitor the transporter localization in membrane subdomain with nanometer resolution (Figure 6.6) [49]. From the single DAT-QD complex trajectory analysis, it was observed that the DAT 615C variant displayed enhanced membrane mobility in comparison to DAT 615R with diffusion rates relative to those observed after lipid raft disruption. Also, single DAT 615C molecules did not exhibit altered mobility in response to cholesterol extraction and amphetamine stimulation, which was quite consistent with the transporter's surface mislocalization and resulted in elevated basal mobility. The increased mobility of the mutant transporter was due to the release of the variant transporter from regulated protein associations. Such approaches can be generalized to future studies to explore the possibilities of disturbed surface DAT dynamics that may occur as a result of genetic alterations and drug usage that may be useful for the treatment of neuropsychiatric disorders.

In another report, biotinylated ligand 2-β-carbomethoxy-3-β-(4-fluorophenyl) tropane (IDT444), which has a high specificity for DAT, has been linked to streptavidin-conjugated QD (SavQD) for the detection of DAT in mammalian cells using flow cytometry and confocal microscopy [50]. Even the low sample preparation utilizing nanomolar and picomolar concentrations of IDT444 and SavQD, respectively, was highly specific for the detection of DAT expression as studied by flow cytometry. The above specific approach was applied for the visualization of acute, PKC-dependent DAT internalization. Such a highly sensitive method based on QDs will allow time-lapse imaging for single DAT molecules in living cells for enhanced time durations when combined with better temporal resolution in order to have a deeper insight to understand molecular dynamics of DAT regulation.

6.6 Limitations of QDs

Despite their potential benefits, QDs have certain drawbacks. Concerns regarding II–IV QD toxicity impede their use in biological and medical applications. Another influential factor affecting QD cytotoxicity is heavy metal ion leakage from the core caused by photolysis and oxidation. To overcome the potential toxicity of QDs caused by cadmium, several approaches have been implemented. Su et al. reported that using both CdS and ZnS shell layers to cap the CdTe core (i.e. core-shell-shell structure) QDs' toxicity reduced, resulting in improved cell viability at high QD concentrations [51]. Cadmium-free InP and CuInS2 have recently emerged as promising candidates for nontoxic QDs. Another disadvantage of QDs for single-molecule imaging is that they blink. To reduce blinking, extremely thick shells of CdS or CdZnS must be developed, and the lattice mismatch between the CdSe core and the shell component must be reduced. As a result, fewer charge carrier trap states exist, and blinking is reduced. CdSe–ZnSe QDs with a size of 8 nm that are completely non-blinking have been reported, with Zn alloyed into the core [7].

6.7 Conclusion

Biomolecules perform a variety of indispensable cell functions including protein synthesis, cell signaling, and DNA transcription. All biological events are either directly or indirectly dependent on molecular interactions. As a result, the ability to detect a single molecule or a small number of molecules in living cells selectively is a very useful tool for understanding the complexities of cellular structures in both animal and plant systems. Some of the other biological applications are rapid DNA sequencing, forensic analysis of biological samples, and medical diagnosis. QD-based single-particle monitoring has allowed us to understand the architecture of plasma membranes and the conformation of synaptic protein complexes. The ligand-conjugated QD labeling strategy has allowed us to unravel the organization of dynamic membranes and also the effect of coding variants associated with brain disorders on the diffusion dynamics of neurotransmitter transporters. The other two major advantages of single-molecule approaches over traditional bulk assays are real-time imaging and a lower sample requirement, which is especially useful when dealing with expensive material.

Fluorescence microscopy in live-cell imaging of dynamic processes generally faces problems in finding the difference in a poor signal against an unstable noise fluctuation from the background, especially due to temporary signal disappearance and heterogeneous molecular motion. Hence the need to develop efficient techniques to reduce image noise, single-molecule detection, and track algorithms for precise and automated analysis of a larger number of single-molecule data. The enormous potential of QDs for single-particle monitoring should be further explored to examine molecular-level dynamics at micro- to-millisecond time intervals in a 3D environment and finally for in vivo analysis. Collaboration among scientists from different fields is required for the broader application of single-molecule monitoring techniques in studying the molecular mechanisms of different biological processes.

References

1 Li, N., Zhao, R., Sun, Y., Ye, Z., He, K., and Fang, X. (2017). Single-molecule imaging and tracking of molecular dynamics in living cells. *Natl Sci Rev* 4 (5): 739–760.

2 Liu, Z., Lavis, L.D., and Betzig, E. (2015). Imaging live-cell dynamics and structure at the single-molecule level. *Mol Cell* 58 (4): 644–659.

3 Xia, T., Li, N., and Fang, X. (2013). Single-molecule fluorescence imaging in living cells. *Annu Rev Phys Chem* 64 (1): 459–480.

4 Kusumi, A., Tsunoyama, T.A., Hirosawa, K.M., Kasai, R.S., and Fujiwara, T.K. (2014). Tracking single molecules at work in living cells. *Nat Chem Biol* 10 (7): 524–532.

5 Chauhan, N., Saxena, K., and Jain, U. (2022). Single molecule detection; from microscopy to sensors. *Int J Biol Macromol* 209: 1389–1401.

6 Pons, T. and Mattoussi, H. (2009). Investigating biological processes at the single molecule level using luminescent quantum dots. *Ann Biomed Eng* 37 (10): 1934–1959.

7 Frasco, M.F. and Chaniotakis, N. (2009). Semiconductor quantum dots in chemical sensors and biosensors. *Sensors (Basel)* 9 (9): 7266–7286.

8 Dabbousi, B.O., Rodriguez-Viejo, J., Mikulec, F.V., Heine, J.R., Mattoussi, H., Ober, R., Jensen, K.F., and Bawendi, M.G. (1997). (CdSe)ZnS core–shell quantum dots: Synthesis and characterization of a size series of highly luminescent nanocrystallites. *J Phys Chem B* 101 (46): 9463–9475.

9 Heyes, C.D. (2019). Chapter 4 – Quantum dots in single molecule spectroscopy. In: *Spectroscopy and Dynamics of Single Molecules* (ed. C.K. Johnson), 163–228. Elsevier.

10 Vasudevan, D., Gaddam, R.R., Trinchi, A., and Cole, I. (2015). Core–shell quantum dots: Properties and applications. *J Alloys Compd* 636: 395–404.

11 Mohamed, W.A.A., El-Gawad, H.A., Mekkey, S., Galal, H., Handal, H., Mousa, H., and Labib, A. (2021). Quantum dots synthetization and future prospect applications. *J Nanotechnol Rev* 10 (1): 1926–1940.

12 Jagtap, S., Chopade, P., Tadepalli, S., Bhalerao, A., and Gosavi, S. (2019). A review on the progress of ZnSe as inorganic scintillator. *Opto-Electron Rev* 27 (1): 90–103.

13 Mashinchian, O., Johari-Ahar, M., Ghaemi, B., Rashidi, M., Barar, J., and Omidi, Y. (2017). Impacts of quantum dots in molecular detection and bioimaging of cancer. *BioImpacts* 4 (3): 149–166.

14 Algar, W.R., Tavares, A.J., and Krull, U.J. (2010). Beyond labels: A review of the application of quantum dots as integrated components of assays, bioprobes, and biosensors utilizing optical transduction. *Anal Chim Acta* 673 (1): 1–25.

15 Lim, S.J., Ma, L., Schleife, A., and Smith, A.M. (2016). Quantum dot surface engineering: Toward inert fluorophores with compact size and bright, stable emission. *Coord Chem Rev* 320–321: 216–237.

16 Haneef, H.F., Zeidell, A.M., and Jurchescu, O.D. (2020). Charge carrier traps in organic semiconductors: A review on the underlying physics and impact on electronic devices. *J Mater Chem C* 8 (3): 759–787.

17 AbouElhamd, A.R., Al-Sallal, K.A., and Hassan, A. (2019). Review of core/shell quantum dots technology integrated into building's glazing. *Energies* 12 (6): 1058.

18 Bera, D., Qian, L., Tseng, T.-K., and Holloway, P.H. (2010). Quantum dots and their multimodal applications: A review. *Materials* 3 (4): 2260–2345.

19 Foubert, A., Beloglazova, N.V., Rajkovic, A., Sas, B., Madder, A., Goryacheva, I.Y., and de Saeger, S. (2016). Bioconjugation of quantum dots: Review & impact on future application. *TrAC Trends Analyt Chem* 83: 31–48.

20 Biju, V., Itoh, T., and Ishikawa, M. (2010). Delivering quantum dots to cells: Bioconjugated quantum dots for targeted and nonspecific extracellular and intracellular imaging. *Chem Soc Rev* 39 (8): 3031–3056.

21 Wu, X., Liu, H., Liu, J., Haley, K.N., Treadway, J.A., Larson, J.P., Ge, N., Peale, F., and Bruchez, M.P. (2003). Immunofluorescent labeling of cancer marker Her2 and other cellular targets with semiconductor quantum dots. *Nat Biotechnol* 21 (1): 41–46.

22 Gao, X., Cui, Y., Levenson, R.M., Chung, L.W.K., and Nie, S. (2004). In vivo cancer targeting and imaging with semiconductor quantum dots. *Nat Biotechnol* 22 (8): 969–976.

23 Guan, Y., Xu, M., Liang, Z., Xu, N., Lu, Z., Han, Q., Zhang, Y., and Zhao, X.S. (2007). Heterogeneous transportation of α1B-adrenoceptor in living cells. *Biophys Chem* 127 (3): 149–154.

24 Gu, Y.-P., Cui, R., Zhang, Z.-L., Xie, Z.-X., and Pang, D.-W. (2012). Ultrasmall near-infrared Ag 2 Se quantum dots with tunable fluorescence for in vivo imaging. *J Am Chem Soc* 134 (1): 79–82.

25 Axelrod, D., Thompson, N.L., and Burghardt, T.P. (1983). Total internal reflection fluorescent microscopy. *J Microsc* 129 (1): 19–28.

26 Mattheyses, A.L., Simon, S.M., and Rappoport, J.Z. (2010). Imaging with total internal reflection fluorescence microscopy for the cell biologist. *J Cell Sci* 123 (21): 3621–3628.

27 Sako, Y., Minoghchi, S., and Yanagida, T. (2000). Single-molecule imaging of EGFR signalling on the surface of living cells. *Nat Cell Biol* 2 (3): 168–172.

28 Chin, C.R., Perreira, J.M., Savidis, G., Portmann, J.M., Aker, A.M., Feeley, E.M., Smith, M.C., and Brass, A.L. (2015). Direct visualization of HIV-1 replication intermediates shows that capsid and CPSF6 modulate HIV-1 intra-nuclear invasion and Integration. *Cell Rep* 13 (8): 1717–1731.

29 Oreopoulos, J., Berman, R., and Browne, M. (2014). Spinning-disk confocal microscopy. In: *Methods in Cell Biology* (ed. Jennifer C. Waters, Torsten Wittman), vol. 123, 153–175. Academic Press.

30 Ma, Y., He, Z., Tan, T., Li, W., Zhang, Z., Song, S., Zhang, X., Hu, Q., Zhou, P., Wu, Y., Zhang, X.-E., and Cui, Z. (2016). Real-time imaging of single HIV-1 disassembly with multicolor viral particles. *ACS Nano* 10 (6): 6273–6282.

31 Tokunaga, M., Imamoto, N., and Sakata-Sogawa, K. (2008). Highly inclined thin illumination enables clear single-molecule imaging in cells. *Nat Methods* 5 (2): 159–161.

32 Zhang, K., Osakada, Y., Vrljic, M., Chen, L., Mudrakola, H.V., and Cui, B. (2010). Single-molecule imaging of NGF axonal transport in microfluidic devices. *Lab on a Chip* 10 (19): 2566.

33 Cui, B., Wu, C., Chen, L., Ramirez, A., Bearer, E.L., Li, W.-P., Mobley, W.C., and Chu, S. (2007). One at a time, live tracking of NGF axonal transport using quantum dots. *Proc Natl Acad Sci* 104 (34): 13666–13671.

34 Ma, J. and Yang, W. (2010). Three-dimensional distribution of transient interactions in the nuclear pore complex obtained from single-molecule snapshots. *Proc Natl Acad Sci* 107 (16): 7305–7310.

35 Rizo, J. (2018). Mechanism of neurotransmitter release coming into focus. *Protein Sci* 27 (8): 1364–1391.

36 Pereda, A.E. (2014). Electrical synapses and their functional interactions with chemical synapses. *Nat Rev Neurosci* 15 (4): 250–263.

37 Huber, K.M., Paoletti, P., and Sjöström, P.J. (2021). Editorial: Latest advances on excitatory synapse biology. *Front Synaptic Neurosci* 13: 50.

38 Moss, S.J. and Smart, T.G. (2001). Constructing inhibitory synapses. *Nat Rev Neurosci* 2 (4): 240–250.

39 Dahan, M., Lévi, S., Luccardini, C., Rostaing, P., Riveau, B., and Triller, A. (2003). Diffusion dynamics of glycine receptors revealed by single-quantum dot tracking. *Science (1979)* 302 (5644): 442–445.

40 Charrier, C., Ehrensperger, M.-V., Dahan, M., Levi, S., and Triller, A. (2006). Cytoskeleton regulation of glycine receptor number at synapses and diffusion in the plasma membrane. *J Neurosci* 26 (33): 8502–8511.

41 Ehrensperger, M.-V., Hanus, C., Vannier, C., Triller, A., and Dahan, M. (2007). Multiple association states between glycine receptors and gephyrin identified by SPT analysis. *Biophys J* 92 (10): 3706–3718.

42 Fagiolini, M., Fritschy, J.-M., Löw, K., Möhler, H., Rudolph, U., and Hensch, T.K. (2004). Specific GABA A circuits for visual cortical plasticity. *Science (1979)* 303 (5664): 1681–1683.

43 Bouzigues, C., Morel, M., Triller, A., and Dahan, M. (2007). Asymmetric redistribution of GABA receptors during GABA gradient sensing by nerve growth cones analyzed by single quantum dot imaging. *Proc Natl Acad Sci* 104 (27): 11251–11256.

44 Varela, J.A., Dupuis, J.P., Etchepare, L., Espana, A., Cognet, L., and Groc, L. (2016). Targeting neurotransmitter receptors with nanoparticles in vivo allows single-molecule tracking in acute brain slices. *Nat Commun* 7 (1): 10947.

45 Wise, R.A. (2004). Dopamine, learning and motivation. *Nat Rev Neurosci* 5 (6): 483–494.

46 Modi, S., Higgs, N.F., Sheehan, D., Griffin, L.D., and Kittler, J.T. (2018). Quantum dot conjugated nanobodies for multiplex imaging of protein dynamics at synapses. *Nanoscale* 10 (21): 10241–10249.

47 Bailey, D.M., Catron, M.A., Kovtun, O., Macdonald, R.L., Zhang, Q., and Rosenthal, S.J. (2018). Single quantum dot tracking reveals serotonin transporter diffusion dynamics are correlated with cholesterol-sensitive threonine 276 phosphorylation status in primary midbrain neurons. *ACS Chem Neurosci* 9 (11): 2534–2541.

48 Chang, J.C., Tomlinson, I.D., Warnement, M.R., Ustione, A., Carneiro, A.M.D., Piston, D.W., Blakely, R.D., and Rosenthal, S.J. (2012). Single molecule analysis of serotonin transporter regulation using antagonist-conjugated quantum dots reveals restricted, p38 MAPK-dependent mobilization underlying uptake activation. *J Neurosci* 32 (26): 8919–8929.

49 Kovtun, O., Sakrikar, D., Tomlinson, I.D., Chang, J.C., Arzeta-Ferrer, X., Blakely, R.D., and Rosenthal, S.J. (2015). Single-quantum-dot tracking reveals altered membrane dynamics of an attention-deficit/hyperactivity-disorder-derived dopamine transporter coding variant. *ACS Chem Neurosci* 6 (4): 526–534.

50 Kovtun, O., Tomlinson, I.D., Sakrikar, D.S., Chang, J.C., Blakely, R.D., and Rosenthal, S.J. (2011). Visualization of the cocaine-sensitive dopamine transporter with ligand-conjugated quantum dots. *ACS Chem Neurosci* 2 (7): 370–378.

51 Sanvicens, N. and Marco, M.P. (2008). Multifunctional nanoparticles – Properties and prospects for their use in human medicine. *Trends Biotechnol* 26 (8): 425–433.

7

Biological Solid-state NMR Spectroscopy

Toshimichi Fujiwara

Institute for Protein Research, Osaka University, Suita, Osaka, Japan

7.1 Introduction

Nuclear magnetic resonance, NMR, is a probe sensitive to the molecular structure and dynamics in solution and solid states [1]. Molecular structures in atomic resolution have been determined by NMR spectroscopy. Solid-state NMR, SS NMR, can provide detailed molecular information for immobilized biomolecules and materials. Typical targets of biological solid-state NMR are proteins, sugars, and lipids in biomembranes and in-situ states such as in cells. Structures of fibrous proteins such as amyloids and crystalline proteins have also been elucidated by SS NMR [2].

There are several high-resolution methods for obtaining detailed structural information on biomolecules for the ultimate determination of their structure. X-ray crystal diffraction, cryoelectron microscopy, EM, and solution and solid-state NMR are among them. The number of protein structures determined by these methods that have been included in the protein databank, PDB, are now about 170,000. These methods for structural analysis require a sample of molecular states depending on the principles of the specific experimental analysis. X-ray diffraction requires quality single crystals to record well resolved diffraction spots. Cryo-EM requires single-particle states having the same structure in vacuum to obtain micrographs of the targeted molecule in all the orientations on the spherical space. These X-ray and cryo-EM experiments are performed at cryo-temperatures to avoid the sample damages due to high-energy X-ray and electron beam irradiation. Solution NMR requires molecules well dissolved in a solution. However, biomolecules are often not amenable to these requirements in studying biological functions. SS NMR has advantages in the sample state: It provides structural information in various states complementary to the states required by the other methods for structural analysis.

Analytical Techniques for the Elucidation of Protein Function, First Edition. Edited by Isao Suetake, Rohit K. Sharma, and Hironobu Hojo.

Nuclear magnetic interactions in solid states are quite different from those in solution states [3]. Static solids have large anisotropic magnetic interactions, which broadens the NMR signals and reduces the spectral resolution. In solution states, however, those orientation-dependent interactions are suppressed due to the rapid thermal random molecular motions, which narrows NMR signals as seen in high-resolution spectra and allows NMR signal assignments to the atoms in the sample molecules. The difference in the molecular dynamics between solution and solid states leads to differences in experimental techniques and instrumentation in NMR. For example, magic-angle spinning of the sample rotor and high-power radio-frequency wave irradiation are used only in SS NMR for obtaining high-resolution spectra [4]. Solid-state magnetic interactions uniquely enable to generate hyperpolarization for nuclear spins by high-field dynamic nuclear polarization, DNP [5].

In this review, the principles of experimental SS NMR are described based on nuclear magnetic interactions. Typical examples of the recent SS NMR applications to biomolecular complexes are also given together with the experimental methods including sample preparation. Readers can find advantages in applying SS NMR in biological problems compared with other methods. One disadvantage of NMR is its low sensitivity compared with X-ray and cryo-EM. This limitation is addressed by enhancement of sensitivity by recent DNP and fast magic-angle spinning under very high magnetic fields, which are also presented along their biological applications.

7.2 Magnetic Interactions for NMR

7.2.1 Zeeman Interaction

Zeeman interaction governs the resonance frequency of nuclear spin, which is proportional to the static magnetic field $\boldsymbol{B}_0$ and is related to the NMR sensitivity. In this review for biological SS NMR, we mainly deal with the application of the spin quantum number $I = 1/2$ with (resonance frequency at 16.4 T, natural abundance): ^{1}H (700 MHz, 100.0%), ^{13}C (176 MHz, 1.1%), ^{15}N (71 MHz, 0.4%), ^{17}F (659 MHz, 100%), and ^{31}P (283 MHz, 100%) because they give narrow resonance lines. Nuclear spins $I \neq 1/2$ (resonance frequency at 16.4 T, natural abundance, I): ^{2}H (75 MHz, 0.01%, 1), ^{14}N (51 MHz, 99.6%, 1), and ^{17}O (95 MHz, 0.04%, 5/2) are also used for biological applications, but the NMR signals are broadened by the large quadrupolar interactions. Rare spins such as ^{13}C and ^{15}N are often enriched in the applications to improve the sensitivity.

7.2.2 Isotropic and Anisotropic Chemical Shifts

The electron cloud surrounding the nucleus shields the nuclear spin from the external static magnetic field and affects the NMR frequency as chemical shifts.

Chemical shifts are much smaller than Zeeman interactions, so they are expressed in ppm unit as a ratio relative to the Zeeman interaction. The electronic state depends on the molecular structure such as covalent bonds, bond lengths and angles, dihedral angles, electrostatic interaction, and hydrogen bonds. Therefore, chemical shifts provide restraints for those structural parameters. Protein secondary structure expressed by main chain torsion angles in a Ramachandran plot can be predicted from the chemical shifts of ^{13}C, ^{15}N, and ^{1}H spins [6, 7]. Generally, nuclei with a larger number of electrons (a larger atomic number) have greater chemical shift dispersion.

The resonance frequencies of the high-resolution NMR lines for solution states provide the isotropic chemical shifts. In static solid states, NMR frequency due to chemical shift is the sum of isotropic chemical shift and anisotropic chemical shift. The anisotropic chemical shift is expressed as a function of the orientation of the atom with respect to the $\boldsymbol{B}_0$ field. The chemical shift anisotropy is suppressed in solutions by the time-dependent fast modulation due to the random rotation over the sphere. The chemical shift anisotropy affects the relaxation of nuclear magnetization and the linewidth of resonance line as a secondary effect in solution states.

The amplitude of anisotropic chemical shift depends on the symmetry of electron wave functions for the spin, e.g. about 30 ppm for methyl ^{13}C and about 150 ppm for ^{13}CO. The distribution width of isotropic chemical shift and typical anisotropy are given in Table 7.1 for ^{1}H, ^{13}C, and ^{15}N. The anisotropic chemical shift gives the information on the molecular structure as well as isotropic chemical shift. The orientation of the atomic group in the molecules can be determined from the orientation dependence of the chemical shift anisotropy. Applications of the chemical shift anisotropy to the structural analysis are given in Section 3.7 on oriented molecular systems.

7.2.3 Homo- and Heteronuclear Dipolar Interactions

Dipole–dipole interactions between spins is termed dipolar coupling. The splitting due to this dipolar coupling is expressed by an equation having the factor $(3\cos^2\theta - 1)/r^3$, where r is the distance between the spins, and θ is the angle between the internuclear vector and the static magnetic field, $\boldsymbol{B}_0$ [3, 4]. Thus, the measurement of the dipolar coupling gives the information on the length and orientation of the internuclear vector, which plays a major role in molecular structural analysis.

Dipolar coupling affects the resonance frequency as the signal splitting in static solid states, while the splitting is eliminated by the random rotation in solutions similarly to chemical shift anisotropy. However, dipolar coupling affects the relaxation of the nuclear magnetization and the linewidth of resonance line as a secondary effect with a factor of $1/r^6$ in both solid and solution states. Typical

Table 7.1 Magnetic interactions for ^{1}H, ^{13}C, and ^{15}N in proteins at a magnetic field of 16.4 T.

Nucleus		Chemical shift/kHz (ppm)		Dipolar couplings/kHz (*r*/nm)		*J* couplings/kHz	
X	Resonance freq./MHz	Shift range	Anisotropy	homo X–X	hetero X–^{1}H	homo X–X	hetero X–^{1}H
^{1}H	700	8 (12)	12 (17, HN)	37 (0.17)	–	0.01	–
^{13}C	176	35 (200)	26 (150, CO)	3.4 (0.15)	23 (0.11)	0.04	0.15
^{15}N	71	11 (150)	14 (200, NH)	0.17 (0.3)	9 (0.11)	0.00	0.09

r: Internuclear distance.

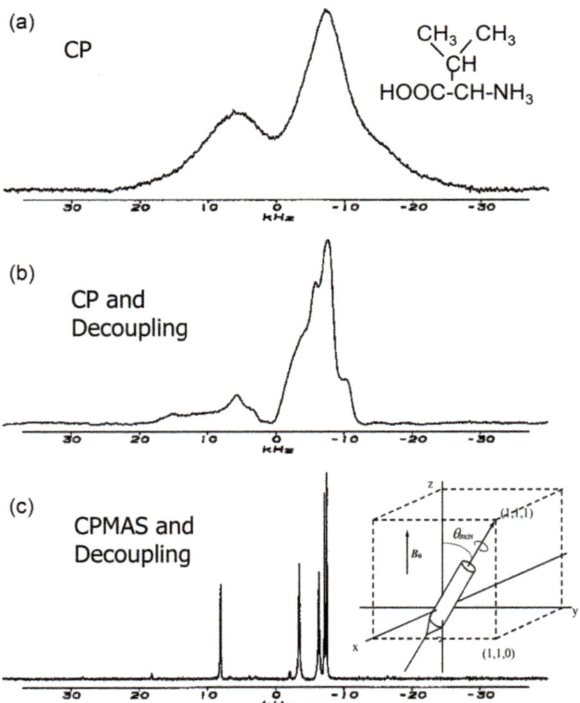

Figure 7.1 High-resolution solid-state NMR spectra for valine at ^{13}C resonance frequency of 100 MHz for (a) static state, (b) static state under CH dipolar decoupling, and (c) spinning sample at the magic angle under the decoupling. The rotation axis is tilted by θ_{MAS} = 54.74° from the static magnetic field.

strengths of homonuclear and heteronuclear dipolar interactions are shown in Table 7.1 for ^{1}H, ^{13}C, and ^{15}N. The dipolar couplings are comparable to chemical shifts in strength. Although J couplings are observable as the splittings in solution states, J couplings are much smaller than dipolar couplings, so they have smaller effects in solid states.

Solid-state NMR gives the spectra in low resolution for powdered static samples where molecules orient randomly over the spherical space (Figure 7.1a). This is because the resonance lines are broadened by the chemical shift anisotropy and dipolar couplings due to their orientation dependence. These broad linewidths are quite different from narrow linewidths of solution NMR, where the chemical shift anisotropy and dipolar couplings are eliminated by Brownian motion. The resonance frequency of solution NMR is primarily determined by isotropic chemical shifts.

7.3 Methods for Solid-state NMR

7.3.1 Sample Preparation of Solid-state NMR

Biological NMR experiments are conducted mainly for $I = 1/2$ nuclei because high-resolution techniques including sample preparation are well established. Carbon-13 and nitrogen-15 are mostly employed for biological NMR despite their low natural abundance. The low abundance nuclei can be enriched for enhancing the NMR sensitivity. The protein samples are uniformly labeled with ^{13}C and ^{15}N by overexpression using *E. coli* cultivation with M9 media whose sole carbon and nitrogen sources are ^{13}C uniformly labeled glucose and $^{15}NH_3Cl$, respectively. The proteins can also be labeled amino-acid-selectively by using the *E. coli* media containing the labeled amino acids.

The sample condition for high-resolution spectra should be optimized so as to reduce the structural distribution. A sample with multiple conformations broadens the resonance lines due to the conformation-dependent isotropic chemical shifts. A crystalline state is one of best sample conditions in this regard, although not all the samples can be crystallized. Internal motion in macromolecules can average the conformation-dependent chemical shift even in solid states, leading to the line narrowing. Low mobilities in low-temperature and lyophilized states often give low-resolution spectra. Thus, the temperature and hydration level of the sample are important parameters for high resolution. Cryoprotectants such as trehalose are often employed for low-temperature SS NMR experiments to form amorphous ice by suppressing crystallization of the water molecules, which disrupts the sample molecular structure.

7.3.2 Experimental NMR Techniques for High-resolution Solid-state NMR

NMR is radio-frequency, RF, spectroscopy. Resonant RF fields are applied to manipulating the nuclear spin states. Mechanical rotation modulates the anisotropic interaction such as chemical shift anisotropies and dipolar interactions due to their orientation dependencies in solids. Therefore, RF field modulation and mechanical sample rotation are used for improving the resolution and sensitivity of SS NMR spectra. High-resolution NMR spectra consist of signals whose resonance frequencies are determined only by isotropic chemical shifts as in solution NMR spectra. These high-resolution solid-state spectra are obtained by decoupling homonuclear dipolar couplings and heteronuclear dipolar and *J* couplings and suppressing the chemical shift anisotropies.

Most high-resolution SS NMR experiments are carried out under heteronuclear RF decoupling and magic-angle spinning, which narrow the linewidths, as shown for ^{13}C NMR of valine in Figure 7.1 [3, 4]. The heteronuclear decoupling is performed by applying a strong RF field to coupled nuclear spins, which are not observed directly. When ^{13}C NMR is observed, proton resonances are irradiated with the RF field to decouple ^{13}C—^{1}H dipolar and *J* couplings. Heteronuclear dipolar couplings are suppressed by random motion in solution states so that only *J* couplings, which are much weaker than the dipolar couplings (Table 7.1), are decoupled in solution NMR. However, in SS NMR, dipolar couplings much larger than *J* couplings must be decoupled. Thus, NMR spectrometers for solids need the generation of RF fields about 100 kHz, which are much stronger than the fields needed for solutions.

Magic-angle spinning, MAS, suppresses anisotropic interactions such as chemical shift anisotropy and dipolar couplings. In ^{13}C-NMR of proteins, MAS eliminates ^{13}C chemical shift anisotropy and ^{13}C—^{13}C, ^{13}C—^{14}N, and residual ^{13}C—^{1}H dipolar couplings under the ^{13}C^{1}H dipolar decoupling. Magic-angle spinning is performed by rotating the sample rotor fast around the axis tilted by the magic angle θ_{MAS} from the static magnetic field, as show in Figure 1c. The high-resolution ^{13}C NMR spectroscopy of organic solids needs about 10 to 20 kHz MAS frequencies at high magnetic fields. The rotor spinning is driven by high-pressure gas with a gas rotor-bearing mechanism. The typical sample rotor is 3 to 4 mm in diameter and made of ceramics such as zirconia to withstand the strong centrifugal force.

Carbon-13 and nitrogen-15 high-resolution spectra of organic molecules are generally obtained by the cross-polarization, CP, method for higher sensitivity. Cross polarization is a double resonance experiment for transferring the ^{1}H polarization to ^{13}C or ^{15}N by the enhanced effect of the heteronuclear dipolar couplings under the strong ^{1}H/^{13}C or ^{1}H/^{15}N double RF resonant fields (Figure 7.2a). The thermal equilibrium ^{1}H polarization amplitude is stronger than the equilibrium ^{13}C and ^{15}N amplitudes by a factor of 4 and 10 in the population difference for the

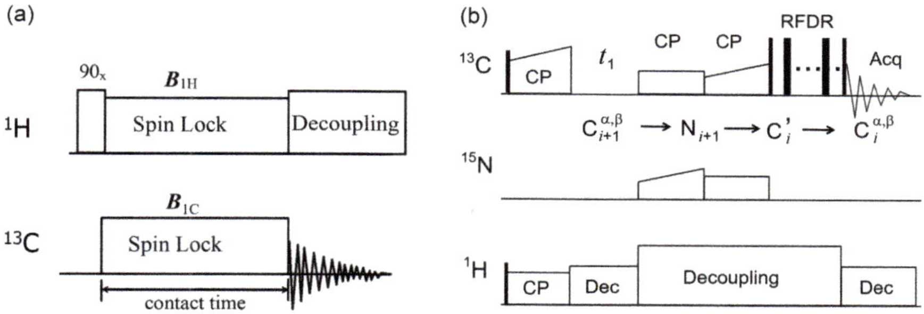

Figure 7.2 RF pulse sequences for solid-state NMR. (a) One-dimensional CPMAS NMR experiment and (b) two-dimensional ^{13}C—^{13}C correlation experiment between $^{13}C^{\alpha}$ in the (i-1)-th amino acid residue and $^{13}C^{\alpha}$ in the i-th residue in protein. The magnetization is transferred successively from the $^{13}C^{\alpha}$ to ^{15}N, from ^{15}N to ^{13}CO, and from ^{13}CO to $^{13}C^{\alpha}$ by the dipolar couplings across chemical bonds in the mixing period.

Table 7.2 Proton and electron polarization depending on temperature and resonance frequency with the wave length at 16.4 T.

	Polarization			
	300 K	**100 K**	**20 K**	**Resonance frequency (Wave length)**
Proton	0.056×10^{-3}	0.168×10^{-3}	0.842×10^{-3}	700 MHz (40 cm)
Electron	0.037	0.110	0.504	461 GHz (0.65 mm)

Polarization: $(N_L - N_H)/(N_L + N_H)$; N_L, N_H: spin populations at the high and low energy states.

two-level systems, respectively, according to their strength of Zeeman interactions (Table 7.1 and footnote for Table 7.2). It is beneficial in the sensitivity to initiate the ^{13}C and ^{15}N NMR experiments with the stronger polarization transferred from the ^{1}H polarization. Additionally, the number of data acquisition per hour of the experiments starting from the excitation of the ^{1}H polarization is larger than the number of those from the excitation of the ^{13}C and ^{15}N polarization. This is because the relaxation time of the ^{1}H polarization is shorter than that of the ^{13}C and ^{15}N polarization. The strong initial polarization and a larger number of data acquisitions make the cross-polarization experiments high sensitivity compared with the single-pulse excitation experiments starting from the ^{13}C and ^{15}N equilibrium polarization. The standard ^{13}C and ^{15}N high-resolution experiments are performed by cross polarization under magic-angle spinning, CPMAS, and the detection under heteronuclear decoupling.

The SS NMR spectrometers need high-power RF amplifiers for ^{1}H decoupling and CP and the mechanism for MAS using large flow of high-pressure gas additionally to solution NMR spectrometers. Thus, dedicated NMR spectrometers are usually employed for high-resolution SS NMR experiments.

7.3.3 Fast MAS for ^{1}H NMR

^{1}H NMR of organic solids are broadened by strong ^{1}H—^{1}H dipolar couplings. It needs a MAS frequency of about 100 kHz to narrow the linewidths by suppressing the homonuclear dipolar couplings [8]. Such fast spinning is performed only by using small rotors whose diameter is less than about 1 mm. The small rotor increases the spinning frequency in inverse proportion to the rotor diameter and enhances the sensitivity per volume by the small receiver coil but reduces the absolute sensitivity due to the smaller sample volume. Thus, small rotors enable the high-resolution and high-sensitivity ^{1}H NMR by the signal narrowing. Higher static magnetic fields are especially beneficial for the resolution of ^{1}H NMR. This is because the ^{1}H linewidths are mainly due to dipolar couplings, so the linewidths in Hz have weak dependence on the static fields whereas the chemical shift differences increase in proportion to the static field strength. This is in contrast with the ^{13}C and ^{15}N NMR linewidths, which are mainly due to distribution of isotropic chemical shifts increasing linearly with the static magnetic field. It is noted that the correlation NMR between ^{1}H and ^{13}C/^{15}N is recorded with higher sensitivity by ^{1}H observation than by ^{13}C/^{15}N observation under the very fast MAS because of the higher resonance frequency of ^{1}H NMR for the detection. Therefore, very fast MAS at very high magnetic fields with ^{1}H detection improves the NMR sensitivity of biological sample such as membrane proteins, which contributes to reducing the sample amount [9, 10]. Deuteration of proteins also reduces the ^{1}H NMR linewidth by weakening the ^{1}H—^{1}H dipolar interactions.

7.3.4 Multidimensional High-resolution NMR Experiments with Recoupling RF Pulse Sequences

The spectral resolution is important for macromolecular structural analysis to increase the number of reliable structural restraints. Spectral resolution can be improved by expanding the spectral space such as in multidimensional NMR spectroscopy. For this purpose, RF pulse sequences for multidimensional NMR have been developed [11]. The spectral axes correspond to the evolution periods for chemical shifts of ^{1}H, ^{13}C, and ^{15}N under MAS.

The evolution periods of the multidimensional NMR pulse sequences are connected by a mixing period of the magnetization by dipolar or *J* couplings (Figure 7.2b). Magic-angle spinning suppresses the anisotropic interactions including dipolar couplings. The effects of dipolar couplings only in the mixing periods are restored by recoupling methods under MAS. Most of the recoupling sequences are RF multiple pulses modulated synchronously with the magic-angle spinning. The design of the pulse sequences depends on the nuclei and MAS frequency.

The structural determination process of NMR consists of the assignments of signals to atoms in the sample molecule and acquisition of internuclear distances from the dipolar couplings. The multidimensional RF pulse sequences are first used for the sequence specific signal assignments where ^{1}H, ^{13}C, and ^{15}N signals are assigned to atoms in amino acid residues on the primary structure of proteins [12, 13]. In these experiments, the mixing periods connect signals from atoms separated by a small number of chemical bonds within a residue and between the residues in dipeptide segments, which respectively provide intra-residue and inter-residue connectivities. The chemical shifts obtained by the signal assignments give backbone dihedral angle constraints for φ and ψ, which specify the secondary structure [7, 14].

The next process is the acquisition of distance constraints. The dipolar mixing period at a longer mixing time provides distance information between atoms up to about 0.8 nm. The signal intensities in these experiments are weaker than those in signal assignment experiments. Thus, higher sensitivities are required for obtaining distance correlation especially for long distances. Semi-selectively isotope labeled proteins are often employed for enhancing the signal sensitivity and for assigning the distance correlations. When the target molecules form a molecular complex as in amyloid fibers, it is also crucial to distinguish between intramolecular and intermolecular distance correlations. The NMR structures are determined from these distance constraints and dihedral angle constraints from chemical shifts. The number of assignments of distance correlations ordinarily increases with the iterative model-building process based on the assignments.

7.3.5 Paramagnetic Effects for Structural Analysis

Unpaired electron spins give much stronger magnetic fields and paramagnetic fields than nuclear spins [15]. The resonance frequencies and magnetic relaxation of nuclear spins are affected by the paramagnetic fields depending on the nuclear-electron distance. Thus, these paramagnetic fields can provide long distances of about a few nanometers, which cannot be accessed by the nuclear–nuclear interactions [16]. The paramagnetic electron can be introduced at specific sites of the targeted molecules by chemical modification such as linkage of S-(1-oxyl-2,2,5,5-tetramethyl-2,5-dihydro-1H-pyrrol-3-yl)methyl methanesulfonothioate, MTSL, having nitroxide radicals to a cysteine residue [15]. In addition to organic radicals, transition metal ions such as Gd^{3+}, Mn^{3+}, Cu^{2+}, and Fe^{3+} are used as the paramagnetic atoms. Some biomolecules such as hem proteins and DNA have endogenous paramagnetic metal ions. The paramagnetic relaxation is also used to reduce the repetition delay. The rapid ^{1}H relaxation propagates to distant protons by the ^{1}H–^{1}H spin diffusion with a rate of about 1 nm/ms^2 in organic solids [4]. The shorter repetition allows the increase in the number of scans, which improves the NMR sensitivity without paramagnetic line broadening.

7.3.6 High-field DNP for Sensitivity Enhancement

The sensitivity of NMR spectroscopy is lower than that of other spectroscopic methods, X-ray diffraction, and cryo-EM because NMR is an RF wave spectroscopy, which relies on the photon energies of RF waves, which are much lower than those of IR, visible and UV lights, X-rays, and accelerated electrons. The sensitivity determines the quality and the upper size limit of applicable biomolecules. Therefore, NMR has been seeking higher resonance frequency in stronger magnetic fields by developing larger superconducting magnets. DNP at high fields is a recently developed method for greatly enhancing NMR polarization for solid-state samples at temperatures lower than about 100 K, as shown in Figure 7.3 [5, 17, 18]. An unpaired electron spin has a Zeeman interaction about 660 times stronger than the ^{1}H spin, as shown in Table 7.2. The ^{1}H polarization for high-resolution NMR experiments is increased by transferring the electron spin polarization to ^{1}H spins using hyperfine couplings under the high field conditions in DNP.

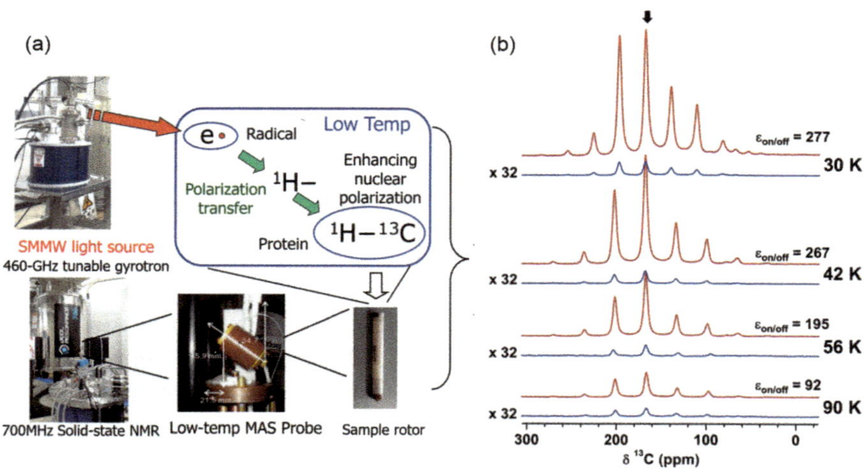

Figure 7.3 DNP-NMR experiments. (a) High-field DNP solid-state NMR spectrometer consisting of a submillimeter wave light source and a low-temperature MAS probe with a high-field NMR magnet, and a sample with unpaired electrons. (b) Enhanced ^{13}C NMR spectra of urea at the ^{13}C resonance frequency of 175 MHz depending on temperature [18], Matsuki et al. (2021) / with permission from Royal Society of Chemistry.

These DNP experiments are initiated by saturating the electron spin polarization at low temperatures. The low temperatures are prerequisite to reduce the electron spin relaxation to enable the efficient polarization transfer by hyperfine couplings. High-power light source for submillimeter-wave, SMMW, irradiation at continuous-wave mode operation is required for the saturation (Figure 7.3). The electron resonance frequency is, for example, about 461 GHz at a static magnetic field of 16.4 T. The 461 GHz wave is an SMMW in the far-infrared region, which is also called terahertz wave. For this purpose, a second-mode gyrotron, a vacuum electron tube, has been developed as an efficient high-power SMMW light source [19].

The MAS rotor is spun by driving and bearing N_2 or He gas depending on the temperature. A closed gas line system for He gas enables DNP–NMR experiments at temperatures lower than about 90 K. This system provides stable gas flow without consuming expensive He gas over a long experimental time [20]. The low-temperature DNP enhancement in sensitivity, S/N, is over 20000 compared with ordinary room-temperature NMR, where DNP enhancement by SMMW irradiation is about 300 in factor, and low temperature effect is about 70 in which the polarization is increased by about 15 and noise is suppressed to 1/5 due to a cold receiver preamp and duplexer. Even without DNP, the He gas low-temperature system increases the sensitivity of the high-resolution SS NMR by a factor of about 70 [21].

The electron spin polarization source for DNP is unpaired electrons in organic radical compounds or transition metal ion complexes including Gd^{3+} and Mn^{2+}. The nuclear magnetization enhancement of protons at high magnetic fields is generally caused by the cross-effect DNP mechanism for two-electron and one-nucleus systems. The matching condition for the DNP is fulfilled when the nuclear Zeeman frequency is equal to the electron resonance frequency difference between the two electron spins. Biradical compounds for this mechanism are optimized as DNP polarization agents, e.g. AMUPol and TOTAPOL. The ^1H polarizations of all the sample molecules within about 10 nm of the agent molecule are enhanced by the spin diffusion from the agent to the sample molecules due to the $^1H—^1H$ dipolar couplings. For the purpose of efficient spin diffusion, the DNP–NMR samples are glass-state solids where the polarization agents are homogeneously mixed with sample molecules. A typical glass solid consist of ^1H_2O:^2H_2O:$^2H_6$-glycerol with a 1:3:6 ratio in volume. The ^1H content in the glass solids is reduced for the optimization of the ^1H spin diffusion without wasting the hyperpolarization on enhancing the proton polarization in the solvent molecules.

7.3.7 Oriented Molecular Systems

Similarly to single crystals, uniaxially oriented samples provide high-resolution NMR spectra with narrow NMR linewidths without using MAS. This resolution is accomplished when the orientation axis is parallel to the static magnetic fields or the sample molecules undergo rapid uniaxial rotational diffusion. The phospholipid membrane sample can be magnetically oriented under the high static magnetic fields owing to the anisotropy of magnetic susceptibility [22]. Typical targets are phospholipids and transmembrane proteins and peptides in the bilayers, which rotate uniaxially along the membrane normal. The magnetic orientation can be monitored by the NMR spectral pattern. Mechanically oriented fibrous molecules such as double helical DNA fibers at high humidity also give narrow signals owing to the uniaxial rotation [23].

Dipolar couplings and chemical shift anisotropies for the oriented molecules provide the information on the angles and dynamics of atomic groups with respect to the orientation axis such as the membrane normal. The two-dimensional $^{15}N^{1}H$ dipolar and ^{15}N chemical shift anisotropy correlation NMR spectra for oriented membrane proteins are known to give a periodic circular cross-peak alignment for an α helical protein segment in accord with the periodicity of the helix [22]. The orientation angle of the membrane normal with respect to the $\boldsymbol{B}_0$ direction can be altered from the parallel to the perpendicular orientation by adding paramagnetic ions such as Yb^{3+}. The structural information obtained from the perpendicular orientation clarifies the ambiguity due to the symmetry in the magnetic interactions. It has also been shown that similar orientational information of atomic groups can be obtained even for powdered membrane proteins and lipids in bilayers owing to their anisotropic rotation in the membranes [24]. Note that the sample orientation experiments in SS NMR employ more highly oriented conditions than those in solution NMR. Thus, stronger dipolar splittings and anisotropic chemical shifts are observed in NMR spectra of solids.

7.4 Applications of Solid-state NMR to Biological Molecular Systems

7.4.1 Membrane Proteins and Peptides

Solid-state NMR provides atomic structural details for sample states not amenable to X-ray crystal analysis and cryo-EM, as mentioned in the introduction. The structure of a membrane peptide tightly bound to phospholipid bilayer membranes has been determined by using internuclear distance and dihedral constraints obtained from SS NMR [13, 14]. This sample system would not form crystals and is too small for cryo-EM structural analysis. Complete signal assignment was performed for uniformly ^{13}C, ^{15}N labeled peptides prepared by overexpression of a protein fused with the peptides using *E. coli* cultivation in minimum media M9. Selectively ^{13}C and ^{15}N labeled peptides were prepared by peptide synthesis for the accurate ^{13}C—^{13}C and ^{13}C—^{15}N distance measurements. These measurements were conducted using recoupling pulse sequences under MAS for high-resolution NMR. The interactions between the peptide and phospholipid bilayers were also studied to determine the peptide molecule location relative to the lipid bilayers, as shown in Figure 7.4a–c [25, 26]. Heteronuclear distance information between ^{31}P at the head groups of lipids and protons in the peptide and that between deuterons labeled in the acyl chains of lipids and protons in the peptide was obtained from the signal intensities for cross polarization by the heteronuclear dipolar interactions. Here, proton polarization components of the peptide were resolved by ^{13}C chemical shifts and were assigned specifically to the residues sequence. This

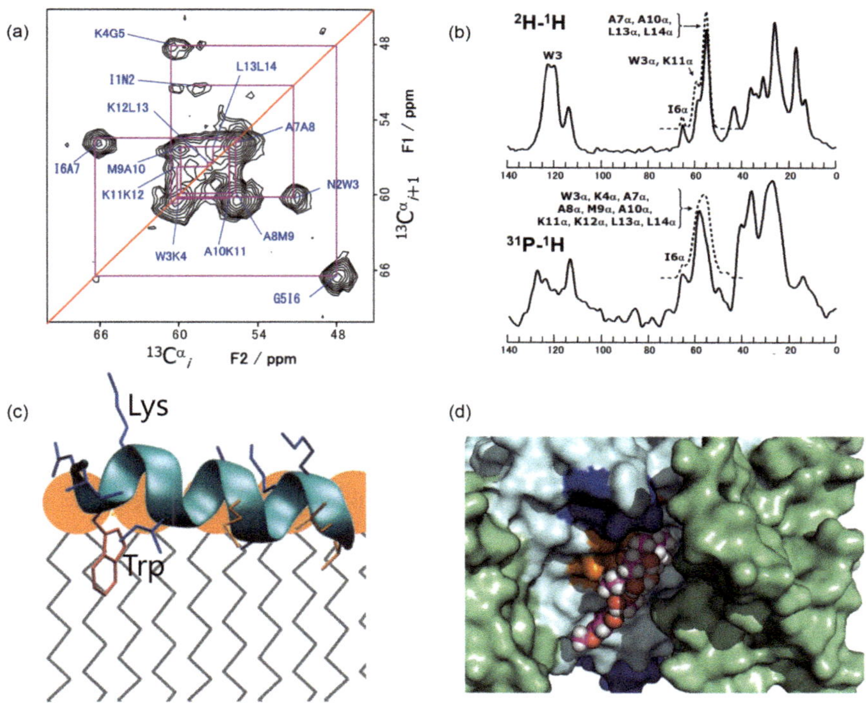

Figure 7.4 (a) Two-dimensional inter-residue $^{13}C^{\alpha}$—$^{13}C^{\alpha}$ correlation spectrum for the sequential signal assignment of mastoparan X, MPX, at the ^{13}C resonance frequency of 125 MHz and the MAS frequency of 12.5 kHz obtained with the pulse sequence given in Figure 7.2b. Adapted from [13]. (b) ^{13}C NMR spectra for ^{1}H-depolarization by intermolecular ^{2}H—^{1}H and ^{31}P—^{1}H dipolar couplings between deuterated phospholipids and MPX. Adapted from [26]. (c) Structural model for MPX-lipid membrane interaction obtained from spectra (b). (d) Structure model showing interaction between veratridine and a helix in Nav 1.4, voltage-gated sodium channel, based on the intermolecular dipolar couplings obtained from SS NMR. Adapted from [27].

analysis revealed the peptide helix was slightly tilted at the membrane interface at the angle of 12° with the tryptophan indole ring inserted into the hydrophobic domain. Similar analysis was performed for the interaction between a partial agonist veratridine and a transmembrane helix of sodium channel Nav1.4 (Figure 4d) [27].

In comparison with distance analysis, secondary structural analysis is not difficult in the sensitivity of SS NMR. The secondary structural analysis from chemical shifts was performed for large protein complexes in transmembrane states, such as *c*-ring of F_oF_1 ATP synthase [28] consisting of 10 subunits having the same structure. SS NMR provides the key information for elucidating the protein functions in lipid bilayers. For example, SS NMR revealed the scorpion toxin–K^+ channel complex structure and its function related to the significant structural changes upon forming the high-affinity binding complex [29]. Atomic structures of closed and open influenza B M2 proton channel obtained from SS NMR revealed the conduction mechanism [30]. Transmembrane allosteric coupling of the gates in a potassium channel was studied by SS NMR [31]. SS NMR and functional assays elucidated conformational changes into an open-activated state upon gating of KirBac1.1 [32]. DNP-sensitivity enhanced SS NMR was also applied to large membrane protein structural studies such as antigenic peptide recognition on the human ABC transporter TAP [33] and photocycle-dependent conformational changes in the proteorhodopsin [34].

7.4.2 Amyloid Fibrous Proteins

Amyloid proteins are related to a group of diseases called amyloidosis. Amyloid-β and α-synuclein are hallmarks of Alzheimer's disease and Parkinson's disease, respectively. The number of patients with these neurodegenerative diseases such as frontotemporal dementia are increasing with the longer life expectancy. Thus, the analysis of the structure of amyloid fibers and its relationship to the function leading to therapeutic medication is crucial in health and medical science and industry. Amyloid proteins form fibrous structures in which the protein molecules assume a unique structure. These proteins do not form crystals so that X-ray crystallography is not applicable. Although cryo-EM provides 3D structures, it often does not provide high-resolution information including side chains. Solid-state NMR can give internuclear distances of about 0.5 nm and backbone dihedral angles that are complementary to the structural information from cryo-EM. Thus, SS NMR is used with many cryo-EM structural analysis of amyloid fibers.

Amyloid-β is a protein segment with about 40 or 42 residues and forms cross-β structure in the fibrils (Figure 7.5a). The structure of the fibril was derived from about 500 $^{13}C—^{13}C$, $^{13}C—^{15}N$ and backbone dihedral angle constraints obtained

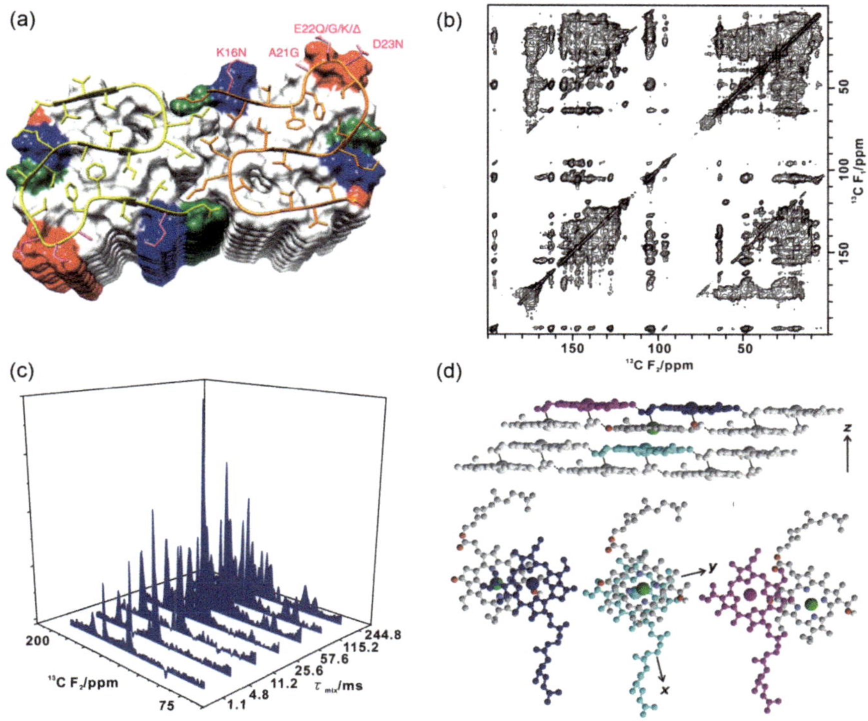

Figure 7.5 (a) Core structure of amyloid β 42 fibril. Hydrophobic (white), polar (green), basic (blue), and acidic (red) residues are shown. [36] / National Academy of Science. (b) ^{13}C—^{13}C spin diffusion spectrum at the mixing time of 115 ms, and (c) the $C3^1$ slices along the F_2 axis at 64 ppm increasing in intensity with the mixing time τ_{mix} for BChl *c* in chlorosomes. These buildups of the signals due to the ^{13}C—^{13}C dipolar interactions gave distance constraints. (d) Structure of light harvesting antenna BChl *c* complex in the cylindrical assembly of chlorosomes. [42] / National Academy of Science. Copyright (2016) National Academy of Sciences for (a).

from multidimensional SS NMR spectra under MAS [35]. The structure of a disease-relevant amyloid-β fibril polymorph was also revealed by SS NMR with cryo-EM [36]. Misfolded α-synuclein with about 140 amino acid residues forms amyloid fibrils. This amyloid structure was determined by SS NMR spectroscopy with validation by EM and X-ray fiber diffraction [37]. This structure had common amyloid features, parallel in-register β-sheets and hydrophobic-core residues. Other amyloid structures such as amyloid protofilaments of β_2-microglobulin fragment of the type I major histocompatibility antigen was studied also by SS NMR [38, 39]. This amyloid fibril precipitates in the kidney of patients receiving hemodialysis for long periods and causes amyloidosis.

7.4.3 In-situ Cellular Biomolecules

Structures of cellular large biomolecular complexes have been studied by SS NMR with advanced technologies for sensitivity enhancement. These biological molecules are immobilized with the intricated structures of molecular complexes in cellular milieu so that SS NMR is a unique method for obtaining the structural information at atomic resolution. Lignin-polysaccharide interactions in plant secondary cell walls was elucidated by SS NMR. Lignin is an aromatic biopolymer and makes plant secondary cell walls waterproof and strong. Lignin also mechanically stabilizes trees and transports water over long distances in xylem. Solid-state NMR revealed that lignin aggregates have hydrophobic nanodomains and interact with xylan at the surface [40]. Peptidoglycan is a polymer consisting of sugars and amino acids that constitutes bacterial cell walls with a mesh-like layer in the cell wall. Bacterial peptidoglycan structure was studied at atomic resolution by proton-detected NMR spectroscopy at 100 kHz MAS frequency without isotope labeling [41]. Structure of light-harvesting bacteriochlorophyll, BChl, *c* assembly in chlorosomes from *Chlorobium limicola* was determined by SS NMR, as shown in Figure 7.5 b–d [42]. Chlorosomes have a cylindrical architecture with a size of about 10 nm with layered BChl molecules. ^{13}C—^{13}C distance measurements for the ^{13}C uniformly labeled BChl complex provided detailed intermolecular interaction sites that elucidated the structure for efficient light harvesting and excitation transfer to the reaction center.

Molecular interactions in biological cells have been studied by solution NMR. However, its applications are limited to small proteins because of viscous and complex cellular environments, which suppress the molecular motion necessary for solution NMR analysis. Magic-angle-spinning SS NMR has an advantage in detecting the signals of molecules under dynamics over a wide range of time constant in cells [43]. The detection is primarily limited by the SS NMR sensitivity. High sensitivity is required to detect proteins especially at their endogenous concentrations. This difficulty can largely be overcome by DNP sensitivity enhancements. These DNP experiments are performed at low temperatures below 100 K for the cells into which the polarization agent was transported. Recently, DNP signal enhancement was applied to NMR detection of proteins inside the mammalian cells by developing cellular DNP techniques [44, 45].

7.5 Concluding Remarks

This review mainly deals with the principles of high-resolution solid-state ^{1}H, ^{13}C, ^{15}N NMR and biological applications based on the current experimental SS NMR methods. Other nuclei, such as ^{2}H and ^{17}O, have large quadrupolar interactions, and ^{19}F has a high resonance frequency comparable to proton with very low abundance in living organisms [30, 46]. We can selectively observe these nuclei at the labeled sites with very small background signals. Oxygen is ubiquitous and an essential element in biology. Thus, the study of ^{17}O in biomolecules should provide important information as well as ^{13}C and ^{15}N, especially for hydrogen bonds. ^{17}O has low sensitivity due to large quadrupolar interaction. However, recent high-sensitivity techniques such as higher magnetic fields about 25 T, very fast MAS, and DNP should partly resolve the difficulties in the sensitivity. Fluorine-19 has advantages of high sensitivity and wide chemical shift dispersion leading to high resolution.

High-field DNP has recently been introduced into biological SS NMR as mentioned in Section 3.6. This method will continue to develop with the sophistication in sample preparation techniques for complex biological systems. Cellular molecules can selectively be detected by optimizing the localization of a polarization agent for DNP enhancement where targeted molecules are selectively labeled with ^{13}C and ^{15}N. The enhancement in the selectivity and sensitivity also needs the optimization of the ^{1}H spin-diffusion pathway with ^{2}H labeling. Very low temperature condition at about 20 K has been employed for further improvements of DNP. Very low temperature also enhances the spin diffusion by reducing the competing process of longitudinal relaxation of the polarization. This allows us to access long-distance structural information on sub-micrometer scales. High sensitivity enables higher dimensionality in NMR as well. Data analysis for such multidimensional NMR spectra has been a time-consuming step in the structural analysis, which requires considerable expertise. This process can be automated by AI deep neural network technology [47].

The unique advantage of SS NMR analysis in structural biology is in the atomic-resolution dynamic structural and interaction analysis of complex systems, which contributes to studying the biological molecular functions under physiological conditions such as in cells. Combination with other complementary methods is quite useful. X-ray analysis and cryo-EM can provide the overall biomolecular structures even though they need crystalline states and single particle states at low temperatures, respectively. Precise structures can be determined by X-ray analysis because of the uniform intermolecular interactions in crystals. Low resolution structures can be obtained easily by EM and optical microscopy. Thus, biological SS NMR is one important method in multi-modal structural analysis based on rapidly developing sophisticated instrumentation with the sample preparation technologies.

References

1 Ernst, R.R., Bodenhausen, G., and Wokaun, A. (1987). *Principles of Nuclear Magnetic Resonance in One and Two Dimensions*. Oxford: Clarendon Press.

2 Reif, B., Ashbrook, S.E., Emsley, L., and Hong, M. (2021). Solid-state NMR spectroscopy. *Nat Rev Methods Primers* 1: Article number 2.

3 Mehring, M. (1983). *Principles of High Resolution NMR in Solids*, 2e. Berlin: Springer-Verlag.

4 Schmidt-Rohr, K. and Spiess, H.W. (1994). *Multidimensional Solid-State NMR and Polymers*. London: Academic Press.

5 Maly, T., Debelouchina, G.T., Bajaj, V.S., Hu, K.-N., Joo, C.-G., Mak–Jurkauskas, M.L., Sirigiri, J.R., van der Wel, P.C.A., Herzfeld, J., Temkin, R.J., and Griffin, R.G. (2008). Dynamic nuclear polarization at high magnetic fields. *J Chem Phys* 128: 052211.

6 Han, B., Liu, Y., Ginzinger, S.W., and Wishart, D.S. (2011). SHIFTX2: Significantly improved protein chemical shift prediction. *J Biomol NMR* 50: 43.

7 Shen, Y., Delaglio, F., Cornilescu, G., and Bax, A. (2009). TALOS+: A hybrid method for predicting protein backbone torsion angles from NMR chemical shifts. *J Biomol NMR* 44: 213–223.

8 Andreas, L.B., Le Marchand, T., Jaudzems, K., and Pintacuda, G. (2015). High-resolution proton-detected NMR of proteins at very fast MAS. *J Magn Reson* 253: 36–49.

9 Agarwal, V., Penzel, S., Szekely, K., Cadalbert, R., Testori, E., Oss, A., Past, J., Samoson, A., Ernst, M., Bçckmann, A., and Meier, B.H. (2014). De Novo 3D structure determination from sub-milligram protein samples by solid-state 100 kHz MAS NMR spectroscopy. *Angew Chem Int Ed* 53: 12253–12256.

10 Lalli, D., Idso, M.N., Andreas, L.B., Hussain, S., Baxter, N., Han, S., Chmelka, B.F., and Pintacuda, G. (2017). Proton-based structural analysis of a heptahelical transmembrane protein in lipid bilayers. *J Am Chem Soc* 139: 13006–13012.

11 Higman, V.A. (2018). Solid-state MAS NMR resonance assignment methods for proteins. *Prog Nucl Magn Reson Spectrosc* 106–107: 37–65.

12 Tanaka, H., Akutsu, H., Yabuta, I., Hara, M., Sugimoto, H., Ikegami, T., Watanabe, T., and Fujiwara, T. (2018). A novel chitin binding mode of the chitin binding domain of chitinase A1 from *Bacillus circulans* WL-12 revealed by solid-state NMR. *FEBS Letters* 592: 3173–3182.

13 Fujiwara, T., Todokoro, Y., Yanagishita, H., Tawarayama, M., Kohno, T., Wakamatsu, K., and Akutsu, H. (2004). Signal assignments and chemical-shift structural analysis of uniformly ^{13}C, ^{15}N-labeled peptide, mastparan X, by multidimensional solid-state NMR under magic-angle spinning. *J Biomol NMR* 28: 311–325.

14 Ikeda, K., Egawa, A., and Fujiwara, T. (2013). Secondary structural analysis of proteins based on ^{13}C chemical shift assignments in unresolved solid-state NMR spectra enhanced by fragmented structure database. *J Biomol NMR* 55: 189–200.

15 Bertini, I., Luchinat, C., Parigi, G., and Ravera, E. (2017). *NMR of Paramagnetic Molecules*, 2e. Amsterdam: Elsevier.

16 Theint, T., Xia, Y., Nadaud, P.S., Mukhopadhyay, D., Schwieters, C.D., Surewicz, K., Surewicz, W.K., and Jaroniec, C.P. (2018). Structural studies of amyloid fibrils by paramagnetic solid-state nuclear magnetic resonance spectroscopy. *J Am Chem Soc* 140: 13161–13166.

17 Thankamony, A.S.L., Wittmann, J.J., Kaushik, M., and Corzilius, B. (2017). Dynamic nuclear polarization for sensitivity enhancement in modern solid-state NMR. *Progress in NMR* 102–103: 120–195.

18 Matsuki, Y., Kobayashi, T., Fukazawa, J., Perras, F., Pruski, M., and Fujiwara, T. (2021). Efficiency analysis of helium-cooled MAS DNP: Case studies of surface-modified nanoparticles and homogeneous small-molecule solutions. *Phys Chem Chem Phys* 23 (8): 4919–4926.

19 Idehara, T., Tatematsu, Y., Khutoryan, E.M., Kuleshov, A.N., Ueda, K., Yamaguchi, Y., Matsuki, Y., and Fujiwara, T. (2015). The development of 460 GHz gyrotrons for 700 MHz DNP-NMR spectroscopy. *J Infrared Millim Terahz Waves* 36: 613–627.

20 Matsuki, Y., Nakamura, S., Fukui, S., Suematsu, H., and Fujiwara, T. (2015). Closed-cycle cold helium magic-angle spinning for sensitivity-enhanced multi-dimensional solid-state NMR. *J Magn Reson* 259: 76–81.

21 Matsuki, Y., Nakamura, S., Hobo, F., Endo, Y., Takahashi, H., Suematsu, H., and Fujiwara, T. (2022). Cryogenic signal amplification combined with helium-temperature MAS DNP toward ultimate NMR sensitivity at high field conditions. *J Magn Reson* 335: 107139.

22 Opella, S.J. and Marassi, F.M. (2004). Structure determination of membrane proteins by NMR spectroscopy. *Chem Rev* 104: 3587–3606.

23 Shindo, H., Fujiwara, T., Akutsu, H., Matsumoto, U., and Kyogoku, Y. (1985). Phosphorus-31 nuclear magnetic resonance of highly oriented DNA Fibers. 1. Static geometry of DNA double helices. *Biochemistry* 24: 887–895.

24 Park, S.H., Das, B.B., Casagrande, F., Tian, Y., Nothnagel, H.J., Chu, M., Kiefer, H., Maier, K., De Angelis, A.A., Marassi, F.M., and Opella, S.J. (2012). Oriented membrane GPCR: Structure of the chemokine receptor CXCR1 in phospholipid bilayers. *Nature* 491: 779.

25 Harada, E., Todokoro, Y., Akutsu, H., and Fujiwara, T. (2006). Detection of peptide–phospholipid interaction sites in bilayer membranes by ^{13}C-NMR spectroscopy: Observation of ^{2}H/^{31}P-selective ^{1}H-depolarization under magic-angle spinning. *J Am Chem Soc* 128: 10654–10655.

26 Ikeda, K., Kameda, T., Harada, E., Akutsu, H., and Fujiwara, T. (2011). Combined use of replica-exchange molecular dynamics and magic-angle-spinning solid-state NMR spectral simulations for determining the structure and orientation of membrane-bound peptide. *J Phys Chem B* 115: 9327–9336.

27 Niitsu, A., Egawa, A., Ikeda, K., Tachibana, K., and Fujiwara, T. (2018). Veratridine binding to a transmembrane helix of sodium channel Nav1.4 determined by solid-state NMR. *Bioorg Med Chem* 26: 5644–5653.

28 Kobayashi, M., Matsuki, Y., Yumen, I., Fujiwara, T., and Akutsu, H. (2016). Signal assignment and secondary structure analysis of a uniformly [^{13}C, ^{15}N]-labeled membrane protein, H+-ATP synthase subunit *c*, by magic-angle spinning solid-state NMR. *J Biomol NMR* 36: 279–293.

29 Lange, A., Giller, K., Hornig, S., Martin-Eauclaire, M.-F., Pongs, O., Becker, S., and Baldus, M. (2006). Toxin- induced conformational changes in a potassium channel revealed by solid- state NMR. *Nature* 440: 959–962.

30 Mandala, V.S., Loftis, A.R., Shcherbakov, A.A., Pentelute, B.L., and Hong, M. (2020). Atomic structures of closed and open influenza B M2 proton channel reveal the conduction mechanism. *Nat Struc Mol Biol* 27: 160–167.

31 Wylie, B.J., Bhate, M.P., and McDermott, A.E. (2014). Transmembrane allosteric coupling of the gates in a potassium channel. *Proc Natl Acad Sci USA* 111: 185–190.

32 Amania, R., Borcika, C.G., Khana, N.H., Versteega, D.B., Yekefallaha, M., Doa, H.Q., Coatsa, H.R., and Wyliea, B.J. (2020). Conformational changes upon gating of KirBac1.1 into an open- activated state revealed by solid- state NMR and functional assays. *Proc Natl Acad Sci USA* 117: 2938–2947.

33 Lehnert, E., Mao, J., Mehdipour, A.R., Hummer, G., Abele, R., Glaubitz, C., and Tampé, R. (2016). Antigenic peptide recognition on the human ABC transporter TAP resolved by DNP-enhanced solid-state NMR spectroscopy. *J Am Chem Soc* 138: 13967–13974.

34 Maciejko, J., Kaur, J., Becker-Baldus, J., and Glaubitz, C. (2019). Photocycle-dependent conformational changes in the proteorhodopsin cross- protomer Asp- His-Trp triad revealed by DNP- enhanced MAS- NMR. *Proc Natl Acad Sci USA* 116: 8342–8349.

35 Colvin, M.T., Silvers, R., Ni, Q.Z., Can, T.V., Sergeyev, I., Rosay, M., Donovan, K.J., Michael, B., Wall, J., Linse, S., and Griffin, R.G. (2016). Atomic resolution structure of monomorphic Aβ42 amyloid fibrils. *J Am Chem Soc* 138: 9663–9674.

36 Wälti, M.A., Ravotti, F., Arai, H., Glabe, C.G., Wall, J.S., Böckmann, A., Güntert, P., Meier, B.H., and Riek, R. (2016). Atomic- resolution structure of a disease-relevant Aβ1–42 amyloid fibril. *Proc Natl Acad Sci USA* 113: E4976–E4984.

37 Tuttle, M.D., Comellas, G., Nieuwkoop, A.J., Covell, D.J., Berthold, D.A., Kloepper, K.D., Courtney, J.M., Kim, J.K., Barclay, A.M., Kendall, A., Wan, W., Stubbs, G., Schwieters, C.D., Lee, V.M.Y., George, J.M., and Rienstra, C.M. (2016). Solid-state NMR structure of a pathogenic fibril of full- length human α-synuclein. *Nat Struct Mol Biol* 23: 409–415.

38 Iwata, K., Fujiwara, T., Matsuki, Y., Akutsu, H., Takahashi, S., Naiki, H., and Goto, Y. (2006). 3D structure of amyloid protofilaments of β_2-microglobulin fragment probed by solid-state NMR. *Proc Natl Acad Sci USA* 103: 18119–18124.

39 Iadanza, M.G., Silvers, R., Boardman, J., Smith, H.I., Karamanos, T.K., Debelouchina, G.T., Su, Y., Griffin, R.G., Ranson, N.A., and Radford, S.E. (2018). The structure of a β 2-microglobulin fibril suggests a molecular basis for its amyloid polymorphism. *Nat Commun* 9 (1): 4517.

40 Kang, X., Kirui, A., Dickwella Widanage, M.C., Mentink-Vigier, F., Cosgrove, D.J., and Wang, T. (2019). Lignin–polysaccharide interactions in plant secondary cell walls revealed by solid- state NMR. *Nat Commun* 10: 347.

41 Bougault, C., Ayala, I., Vollmer, W., Simorre, J.-P., and Schanda, P. (2019). Studying intact bacterial peptidoglycan by proton- detected NMR spectroscopy at 100 kHz MAS frequency. *J Struct Biol* 206: 66–72.
42 Egawa, A., Fujiwara, T., Mizoguchi, T., Kakitani, Y., Koyama, Y., and Akutsu, H. (2007). Structure of the light-harvesting bacteriochlorophyll *c* assembly in chlorosomes from *Chlorobium Limicola* determined by solid-state NMR. *Proc Natl Acad Sci USA* 104: 790–795.
43 Luo, Y., Xiang, S., Hooikaas, P.J., van Bezouwen, L., Jijumon, A.S., Janke, C., Förster, F., Akhmanova, A., and Baldus, M. (2020). Direct observation of dynamic protein interactions involving human microtubules using solid-state NMR spectroscopy. *Nat Commun* 11: Article number 18.
44 Ghosh, R., Xiao, Y., Kragelj, J., and Frederick, K.K. (2021). In-cell sensitivity-enhanced NMR of intact viable mammalian cells. *J Am Chem Soc* 143 (44): 18454–18466.
45 Narasimhan, S., Scherpe, S., Paioni, A.L., van der Zwan, J., Folkers, G.E., Ovaa, H., and Baldus, M. (2019). DNP-supported solid-state NMR spectroscopy of proteins inside mammalian cells. *Angew Chem Int Ed* 58: 12969–12973.
46 Wu, G. (2019). 17O NMR studies of organic and biological molecules in aqueous solution and in the solid state. *Progress in NMR* 114–115: 135–191.
47 Kobayashi, N., Hattori, Y., Nagata, T., Shinya, S., Güntert, P., Kojima, C., and Fujiwara, T. (2018). Noise peak filtering in multi-dimensional NMR spectra using convolutional neural networks. *Bioinformatics* 34: 4300–4301.

8

Electrically Induced Bubble Knife and Its Applications

Yoko Yamanishi

Department of Mechanical Engineering, Kyushu University, Fukuoka, Japan

8.1 Introduction

"Microbubbles" ranging in size from a few micrometers to several tens of a micrometer have been attractive for biomedical applications such as drug delivery and therapeutic ultrasound. For the biomedical field in particular, bubbles a few micrometers in size are useful in terms of the transportation, reaction to, and attack of limited regions for minimally invasive treatment, so the production of mono-dispersed bubbles a few micrometers in size is in high demand [1, 2]. Conventionally, several methods of producing microbubbles have been investigated, such as generation in a microfluidic T-junction [2], the flow-focusing method [3], and chopping by actuators [4], but producing stable mono-dispersed microbubbles a few micrometers in size in sufficient quantities remains a challenge. On the other hand, recent progress in the area of electrically induced fluids have provided many applications, such as a microjet driven by electro-conjugate fluid [5], chemical ionization, and deposition by electrospray techniques [6, 7]. Electrosprays can atomize reagent to droplets ranging in size from nanometers to micrometers, but this technique is not applicable to bubble generation.

This chapter describes a micro-electric knife overlapped by an insulation glass layer with a novel structure, which we term a "bubble-reservoir," to produce stable, mono-dispersed microbubbles and target cell membrane ablation using the generated microbubbles as a biomedical application. Recently, manipulation and processing technologies for microorganisms and cells have been one of the important subjects for the development of micro/nano processing technology, gene technology, and neurology [8–11]. However, they require special environments, such as in a medium (water) and atmospheric pressure, which have limited the devices that can be operated. For example, electron beam machining (EB) [12] or focused ion beam machining (FIB) [13], which are good for nanoscale

Analytical Techniques for the Elucidation of Protein Function, First Edition. Edited by Isao Suetake, Rohit K. Sharma, and Hironobu Hojo.

fabrication, cannot be used in the atmospheric environment. The use of a femtosecond laser for cell processing has recently become a promising application. However, such devices tend to be huge in size and high in cost, and they take a long time to process a wide area. Instead, an electric knife is a good candidate for cell processing. Palanker et al. have reported micrometer-order ablation by using a 12 μm metal blade with SiO_2 on a fresh porcine cornea in vitro [14–17]. However, there was thermal collateral damage. The micro-electric knife is a highly accurate and minimally invasive operation using the crush of the microbubbles generated by the micro-electric knife without thermal collateral damage. For example, this technique can contribute to the reduction of the removed area of oocyte in enucleation, improving the production rate.

8.2 Electrically Induced Bubble Knife

The electrically induced bubble knife was initially proposed for robust usages in wet environments for biomedical applications as a micro-electric knife. However, micro-electric knife corrosion tends to occur when the size of the electrode is very small. Moreover, electrical discharge in the medium (electrolyte solution) has several inherent problems, such as the generation of undesired and uncontrollable bubbles due to heat and electrolysis as well as electrode contamination due to the protein adhesion when it is used for biological materials (Figure 8.1). After several preliminary experiments, it was confirmed that the commercial coating is not sufficient to insulate the electrode perfectly, and we have decided to control bubbles and utilize them for local ablation of a single cell. To produce a perfectly insulated micro-electrode, a glass capillary tube (50 μL; Drummond Scientific Co.), a copper micro-wire 30 μm in diameter, and a silver paste were used. The process flow for the fabrication of an electrode using the glass puller with a copper micro-wire is shown in Figure 8.2. In this process, the insulation layer and a wire were fabricated simultaneously to produce a knife for perfect insulation. First of all, a copper micro-wire was inserted into the glass tube. Then, the glass puller (P-1000 IVF; Sutter Co.) was employed to pull the glass tube using the joule heat of a

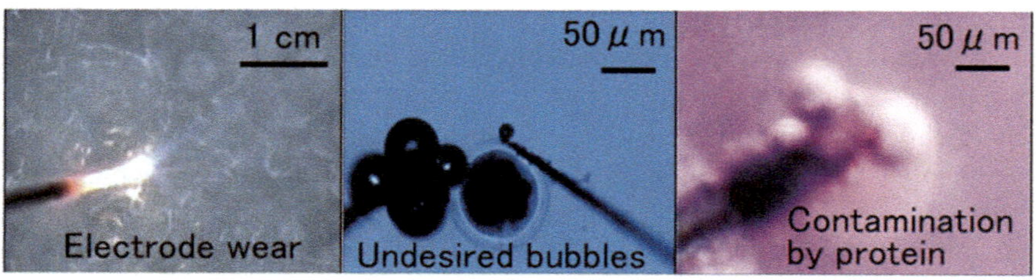

Figure 8.1 Problems of electrical ablation in water environment.

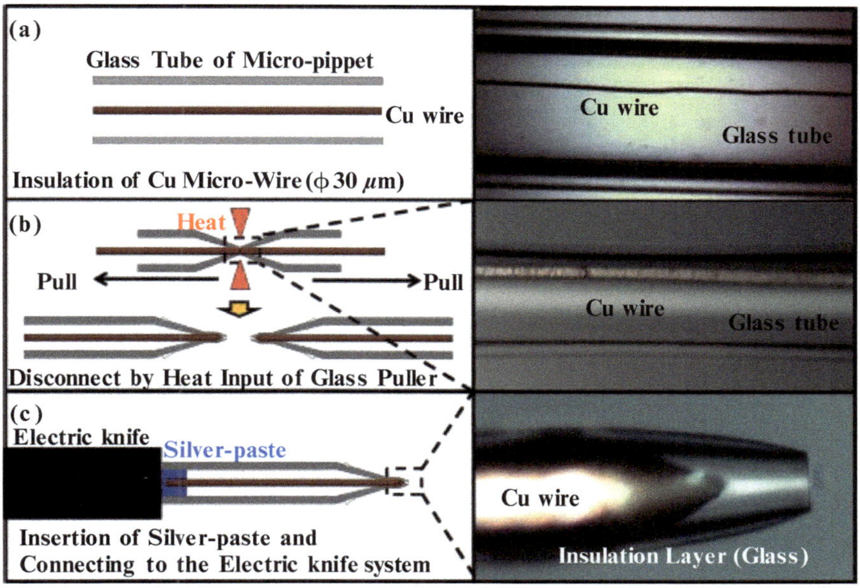

Figure 8.2 Process flow to fabricate an electrically induced bubble knife.

resistance unit. After the programmed thermal input was applied, the tube and the wire were disconnected. Finally, the silver paste was stuffed into the end of the tube to connect it to the power supply of the general electric knife. Due to the viscoelasticity of glass insulation, glass can be extended further than copper, producing a tapered space we term a "bubble reservoir," which stabilizes the electric discharge and bubble generation. The shape of this structure can be controlled by programming the thermal input. Figure 8.3 shows the characteristics of the tip of the electrically induced bubble knife. Its x-axis is the tensile force for pulling the glass tube (defined as pull, the function of the applied current of the glass puller). O is the origin. In this figure, we found that the more tensile force is added to the glass tube, the longer the bubble reservoir (L) becomes and the smaller the diameter of the edge of the bubble reservoir (D) becomes. However, when pull is more than 70, that tendency becomes erratic. This is due to the fact that the glass tube cannot follow up the pull speed due to the viscoelasticity of borosilicate glass. Figure 8.4 shows the effectiveness of the bubble reservoir, which was mounted at the tip of the glass electrode. It was observed that directional, mono-dispersed microbubbles line up when discharged by the electrically induced bubble knife. However, when the electrically induced bubble knife is used without the bubble reservoir, it was observed that scattered poly-dispersed microbubbles were produced. The electrical circuit is shown in Figure 8.5. Non-inductive resistance (10.8 kΩ) was added to the commercial electric knife power supply (Hyfrecator 2000;

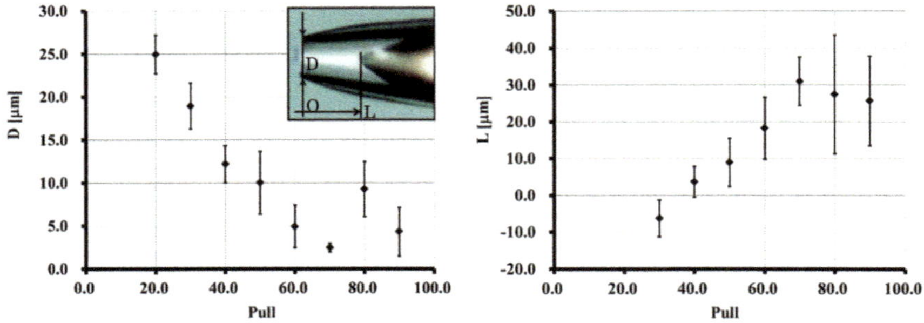

Figure 8.3 Characteristics of micro-electrode (error bar denotes the standard deviation; $n = 6$). *Pull is the unit that is the function of the applied current to the electromagnetic coil controlling the pulling speed.

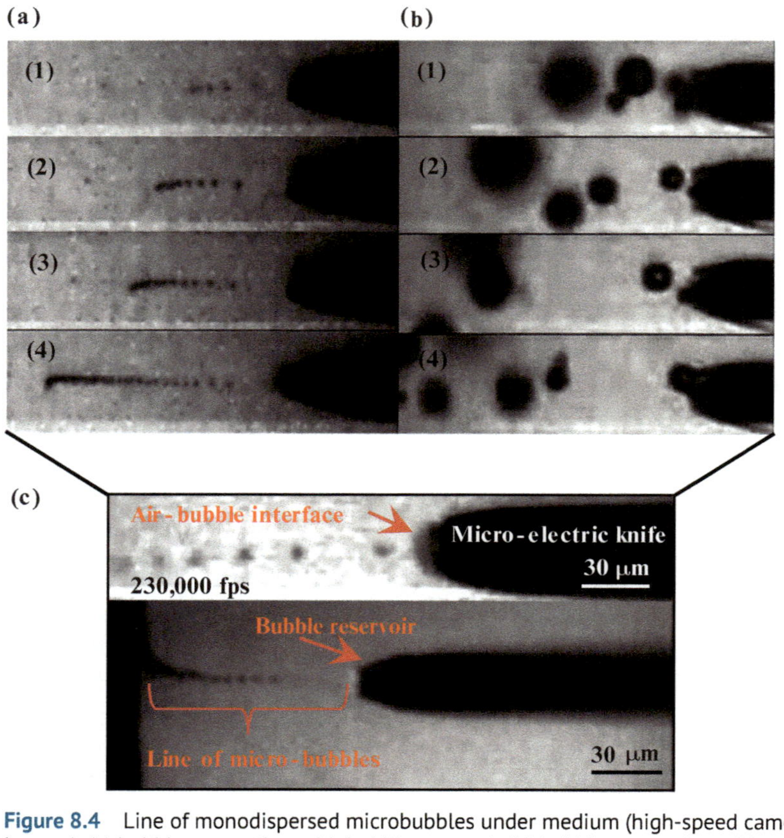

Figure 8.4 Line of monodispersed microbubbles under medium (high-speed camera images): (a) bubble generation with bubble reservoir, (b) bubble generation without bubble reservoir, (c) magnified view of line of monodispersed microbubbles and bubble reservoir.

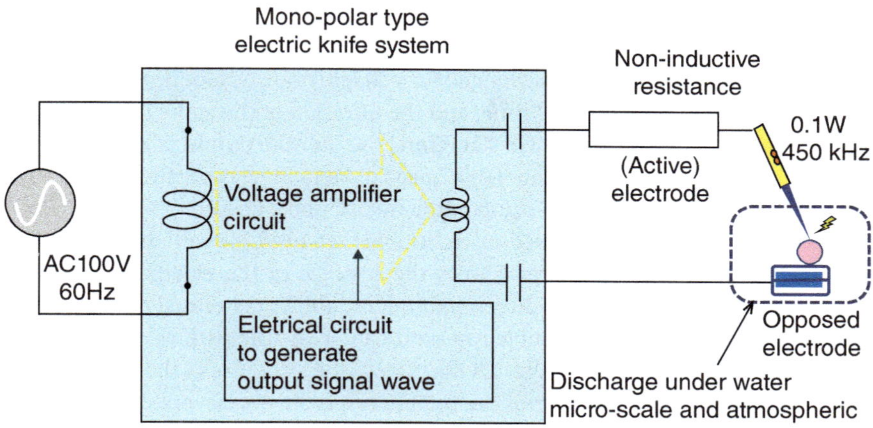

Figure 8.5 Overview of electrical circuit of electrically induced bubble knife.

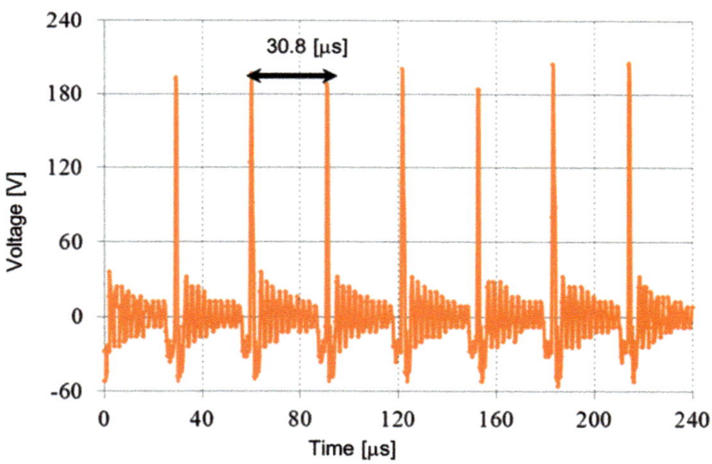

Figure 8.6 Input voltage for microbubble knife as a function of time.

Kobayashi Medical) to adapt the output power to the cellular-scale ablation. There is electrolyte solution between the active electrode and the opposed electrode. For the current study, cell culture medium 199 (Thermo Fisher Scientific K. K.) was used to simulate a wet environment for biological applications. The input voltage to the system is shown in Figure 8.6. Two condensers shown in Figure 8.5 were placed to prevent any direct current produced by the rectification between the active electrode and the biological object that was designed for the conventional electric knife. A sharp spike in the voltage, about 200 V, was applied periodically to the electrically induced bubble knife. The interval between applications of the voltage was about 30.8 μs (small vibration between spikes is for electrical stability). Also, each spike profile was investigated for voltage (orange color line) and current (red color line) as a function of time, and the duration of the spike in voltage was only 1.6 μs as shown in Figure 8.7. Therefore, the thermal output of the signal was calculated to be 2.7 μJ, very small. High-speed camera images confirmed that each spike corresponded to the expansion of the microbubble (Figure 8.4).

The generated mono-dispersed microbubbles are used for cell ablation as biomedical applications. Figure 8.8 shows the concept of the electrically induced bubble knife. Mono-dispersed directional microbubbles generated by the oscillation of microbubbles in the bubble reservoir can ablate the surface of a cell using the cavitation phenomenon. Figure 8.8(a) shows the structure of the bubble reservoir in detail. The tip diameter of the bubble reservoir was several times the distance between the tip of the active electrode and the tip of the probe. Figure 8.8(b) shows the sequence process of the bubble crush when it is close to the target object. First of all, a single bubble is produced by electrolysis or local heat due to the electric discharge, and then the mono-dispersed directional bubbles are created by fluidic oscillation. Finally, cell ablation can be achieved by the crush of high-speed bubbles in a process known as cavitation. A recent paper confirmed that the resolution of the micro-jet when a bubble is crushed can be down to the nanometer order [18]; hence, it is expected that the resolution of ablation can be down to the nanometer order as well.

The ablation of zona pellucida, which is a glycoprotein layer surrounding the plasma membrane of mammalian **oocytes,** was carried out to evaluate the

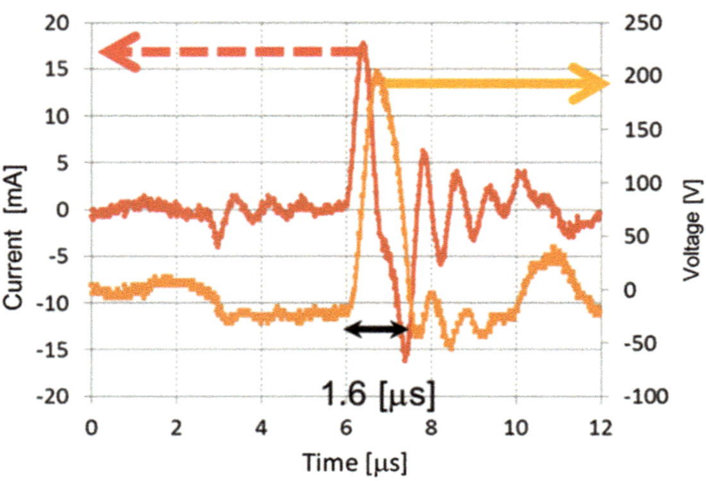

Figure 8.7 Profiles of current and voltage for a single-pulse discharge.

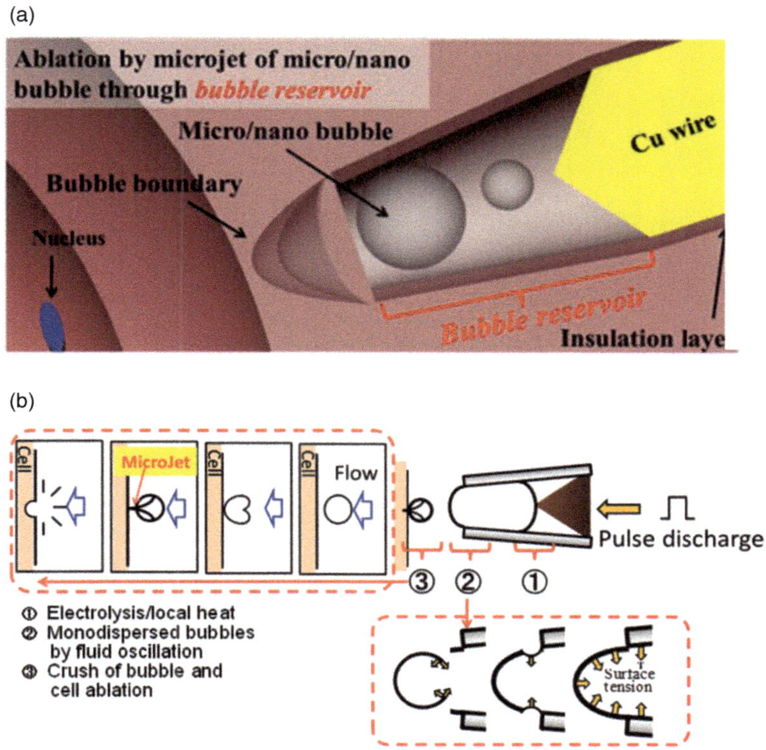

Figure 8.8 Concept of cell ablation by micro/nano bubbles generated by the electric knife: (a) concept view of micro-electric bubble knife and bubble reservoir, (b) mechanism of ablation.

ablation ability of the electrically induced bubble knife. A bovine oocyte was used as the biological target. To clarify the region of the ablation, the oocyte was fluorescently dyed with Rhodamine B, and the cross section was observed with a confocal microscope. The sequence of images shows the ablation of the oocyte by the crush of microbubbles. Figure 8.9(a) shows the ablation of the oocyte when the electrode is in contact with the surface of the oocyte ($D = 0$ μm in Figure 8.10). Figure 8.9(b) shows the ablation when the electrode is separated from the oocyte. Figure 8.10 shows the relationship between the ablation width (w) as a function of the distance (D) from the microelectrode to the zona pellucida. It was observed that the resolution of this ablation reached a size smaller than the tip of the electrically induced bubble knife, whose diameter was 10.2 μm. The dotted line in Figure 8.10 shows the diameter of the electrically induced bubble knife. The resolution of the zona pellucida ablation can be on the order of a few micrometers.

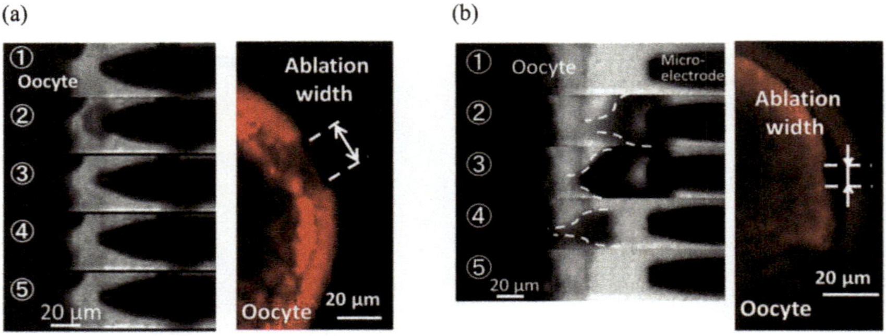

Figure 8.9 Ablation of cell membrane: (a) bubble knife is in contact with the cell, (b) bubble knife is off the cell (zona pellucida and cytoplasm was dyed with Rhodamine B, and the picture was taken with a confocal microscope).

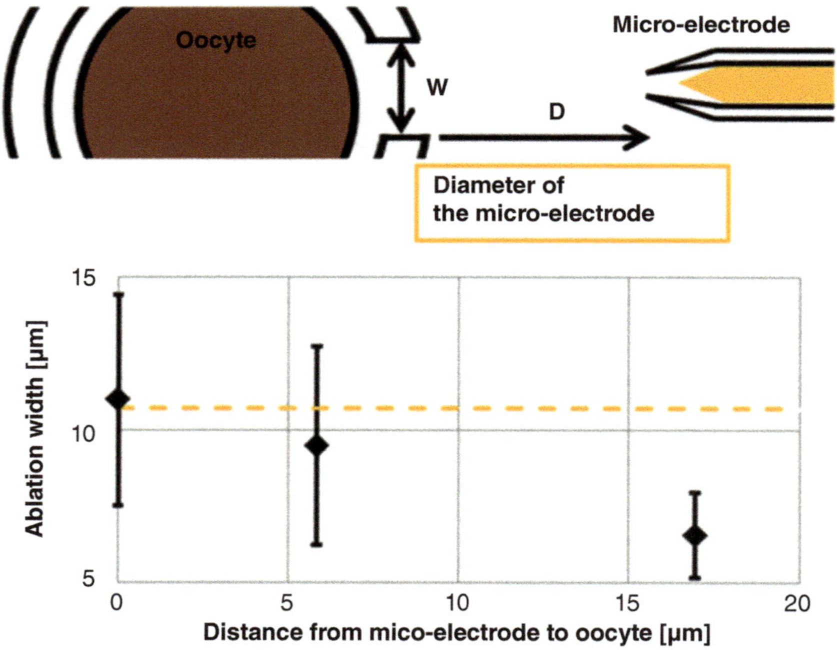

Figure 8.10 Ablation of cell membrane (a) bubble knife is in contact with the cell; (b) bubble knife is off the cell (zona pellucida and cytoplasm was dyed by Rhodamine B, and the picture was taken by confocal microscope) ($N = 4$).

8.3 Electrically Induced Bubble Injector

Figure 8.11 shows the concept of the electrically induced microbubble knife and injection system. The reagent phase was introduced through the outer microchannel of the probe by using a syringe pump. The inner side of the probe has an electrode and a space of the bubble reservoir to generate stable electrically induced bubbles. The electrical circuit used for the system was modified from the conventional electric knife to adapt to cell ablation. The output voltage was amplified in the general electric knife circuit. A non-inductive resistance, which was installed to the circuit to adapt to cellular ablation, was 10.82 kΩ. Discharging time and timing was controlled by a digital input-output board in which the time resolution was 1 ms. The bubble reservoir has characteristics to generate directional mono-dispersed microbubbles, and the bubbles located in the bubble reservoir can make an ablation of the cell surface when they crushed on it due to the cavitation. This enables to make a cellular scale ablation because the tip diameter of the bubble reservoir is a few micrometers, and we utilized this cavitation for the enucleation of bovine oocyte [19]. In that experiment, the cavitation made ablation of zona pellucida of bovine oocyte whose hardness was about ten times larger than cytomembrane; thus the cavitation was used as a powerful injection method. At the same time, directionally dispersed mono-dispersed microbubbles from the bubble reservoir can transport an exogenous gene or a reagent toward the nucleus by electrically induced fluidic force with the aid of the surface tension and capillary force at the air-liquid interface. Because of these configurations, microbubbles enclose the reagent and exogenous gene in the air-liquid interface. Once microbubbles are dispensed from the outlet, a flow that transports the reagent solution is generated, and microbubbles can assists the injection event.

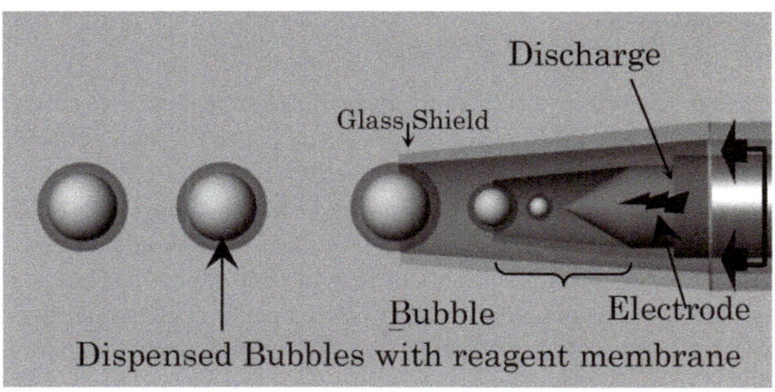

Figure 8.11 The concept of the micro-electric bubble knife and injection mechanism. Microbubbles are dispensed with reagent membrane by electric discharge and reagent inlet.

8.3.1 Bubble Formation with Reagent Interface

This section describes the transportation ability of an air-liquid membrane of plasma-induced bubbles. The first trial of the dispensing of the bubbles with regent surface was the methylene blue (Wako Pure Chemical Industries, Ltd). It was observed that the blue color around a bubble was maintained at a distance more than 100 μm from the tip of the electrode. This means that the diffusion of methylene blue solution to the ambient solution (medium) is much slower than the speed of dispensing bubbles, which is in the order of milliseconds. The dispensing bubble size can be calculated by capillary wave based on surface tension [1]. The air-liquid interface vibration induced by pressure wave of microbubble generation at the top of the copper wire by electric discharge is in the ultrasonic range (the frequency of electric pulse is about 30.8 kHz, which is linked to the bubble expansion in the bubble reservoir and width of electric pulse only about 1.6 μs). The capillary wavelength can be obtained from the well-known Kelvin equation [20] as follows:

$$\lambda = (8\pi\sigma / \rho f^2)^{1/3}. \tag{8.1}$$

In this equation, λ is the capillary wavelength. σ is surface tension. ρ is the density of the solution, and f is the applied frequency to the electric circuit. The fragmentation mechanism of bubbles in an ultrasonic field, which is controlled by inertia, viscosity, and surface tension, can be explained by the capillary hypothesis. The average size (D) of the bubbles fragmentized by an ultrasonic wave can be $D = 0.34\lambda$ or smaller [21]. For the present study the calculated bubble size estimated was less than 18.8 μm. On the other hand, the diameter of dispersed microbubbles was smaller than the calculation in most of the cases. It seems that other physical factors have to be taken into consideration, which requires further research.

8.3.2 Simultaneous Injection and Ablation

Finally, simultaneous ablation of a cell and introduction of a reagent was carried out for a bovine oocyte. A bovine oocyte, which is composed of zona pellucida (transparent shell) and cytoplasm, was used for the experiment of simultaneous ablation and injection. Generally, Young's modulus, which is a measure of elasticity that shows how a substance stretches and/or deforms when under stress, is ten times larger for the zona pellucida than for the cytomembrane; therefore, it is reasonable to assume the zona pellucida as cell wall of the plant cell (its hardness varies among species, but it is harder than the cytomembrane, and electroporation can't ablate it). Figure 8.12 shows the experiment to determine whether the electric discharge power is enough to make an ablation to the cell wall simply.

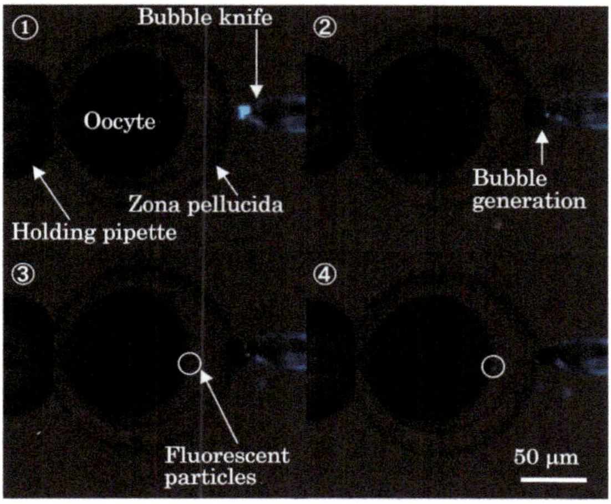

Figure 8.12 Ablation and injection of bovine oocyte. Bovine oocyte was fixed by holding pipette. Blue fluorescing substance was fluorescent particles whose diameter was 100 nm.

In this figure, the blue object was fluorescent particles whose diameter is 100 nm. The fluorescent particle suspended solution was filled in the loading chemical of the bubble knife. From this experiment, we confirmed bubble generation and fluorescent particles existed in the zona pellucida of the bovine oocyte. Finally the simultaneous ablation and injection was successfully operated.

8.4 Plasma-induced Bubble Injector

Figure 8.13 shows the concept of bubble and plasma injector. When the electric pulse is applied to the microelectrode, microbubbles are generated from its tip, which is called "bubble reservoir." Immediately after the generated microbubble was collapsed, an ultrafine high-speed liquid flow, like a liquid needle, called "micro-jet" [22] penetrates the bubble and perforates soft materials such as animal cells with transporting reagent (Figure 8.13(a)). According to past studies, the air-liquid interface of microbubbles has adsorption force by electrical charge on the interface [23], and microbubbles (whose diameter is less than 50 μm) shrink and disappear with certain amount of time [24]. These characteristics also contribute to reagent transportation in liquid phase and minimal invasiveness, respectively.

For the current study, the synergistic effect of cavitation of bubble and plasma ablation was used for hard materials (Figure 8.13(b)).

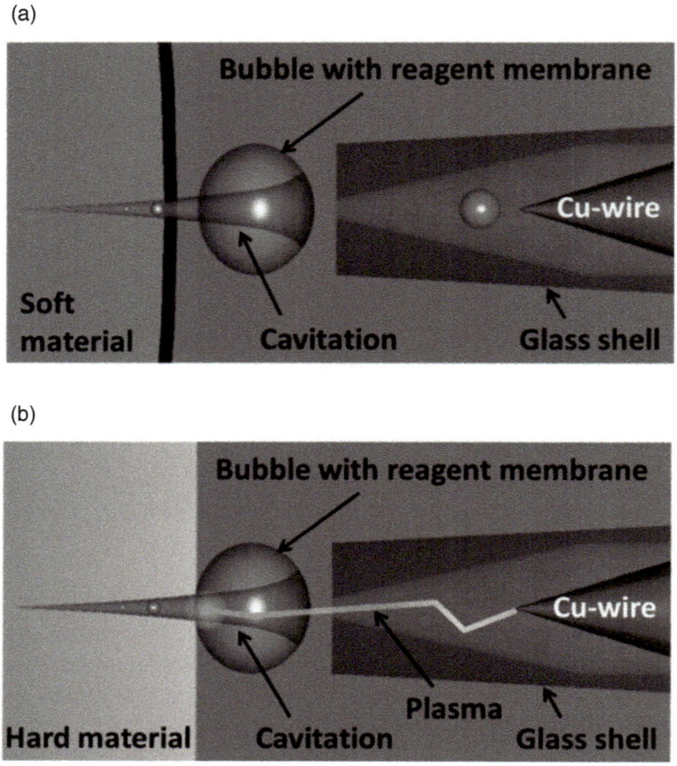

Figure 8.13 Concept of bubble-plasma injector: (a) injection using bubble cavitation for soft material, (b) injection using plasma and bubble cavitation to hard material.

8.5 Protein Crystallization by Electrically Induced Bubbles

Conventionally, drug development tends to be high in cost and time consuming because good medicines development can only be achieved through many trials and errors. In order to solve such issues, researches on protein crystallization tend to draw attention especially in the analysis of the protein structure for smart drug development [25]. To decide the structure of a protein, it is promising to produce a protein crystal and to analyze it using X-rays. For example, one of the conventional protein crystallization methods is the hanging drop method and the sitting drop method, which use supersaturation of the protein solution and are classified as vapor diffusion [26, 27]. However, the method tends to be time consuming, although it produces good quality of the protein crystal. A femtosecond laser has recently been proposed to produce protein crystallization [28]; however, this tends to have a high in cost, as shown in Table 8.1.

Table 8.1 Classification of the protein crystallization method.

	Vapor diffusion	Femto-second laser	Bubble knife
Speed	Low	High	High
Efficiency	Low	High	High
Cost	Low	High	Low

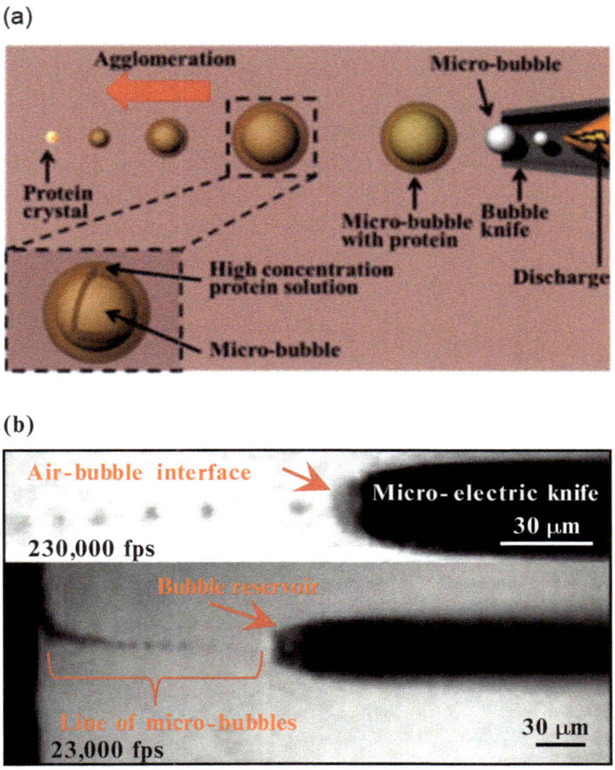

Figure 8.14 (a) Concept of the protein crystallization by electrically induced bubble knife. Microbubble with protein solution dissolved and protein agglutinated by microbubble; (b) underwater discharging by bubble knife. Electrically induced microbubble whose generation speed is 30.8 kHz made line and negative pressure flow.

Figure 8.14 shows the proposed new method of protein crystallization. The electrically driven microbubbles were dispensed and coated with protein molecules by their stiction force on the surface of bubbles. Next, the microbubble shrank and protein concentration on the bubble was increased because the superficial area of the bubble was decreased. Finally, at the time of reaching protein concentration to supersaturation of the protein solution, a protein crystal core was generated.

This phenomenon can occur in low concentration of protein solution, and this is important to produce a protein crystal of good quality [29]. Figure 8.15 shows electrically driven mono-dispersed microbubbles in underwater discharging in a medium. This mono-dispersed bubble flow enables to involve ambient solution and enhances the interfusion of microbubbles and protein. Electrically induced microbubble knife (bubble knife) was produced by the following three steps.

First of all, a Cu wire was installed in the glass tube. Next, the Cu wire and glass tube were pulled on heating by glass puller simultaneously. During this time, the glass was extended longer than the Cu wire by the difference of viscoelasticities. Consequently, there is a region in the glass insulation layer at the tip of the Cu wire, which was termed bubble reservoir; this reservoir contributes stabilization for the production of microbubbles [30]. Finally, a bubble knife was connected to an electrical circuit by the silver paste. The experimental setup and experiment procedure are shown in Figure 8.15. The discharging time and timing of the

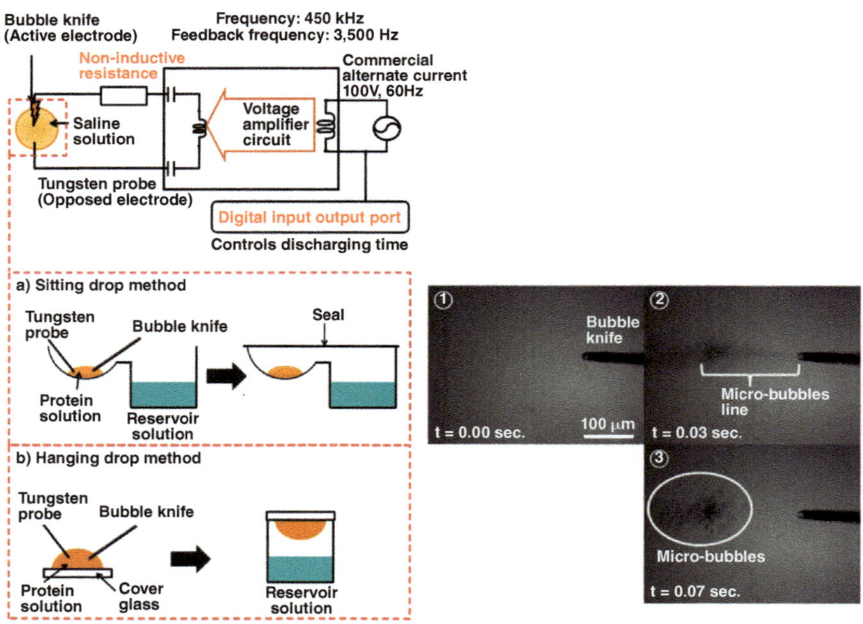

Figure 8.15 Electrical circuit of experimental setup and experiment procedure and discharging bubble knife in the protein solution.

electrically driven bubble knife were controlled by a digital input/output port. The sitting drop method and the hanging drop method were employed. In the hanging or droplet method, the bubble knife was discharged in a protein solution on a drop well. After discharging, the drop well and reservoir well were sealed. In the hanging drop method, the bubble knife was discharged in a protein solution on the cover glass. After discharging, the reservoir well was sealed by the cover glass. The experimental conditions of each well are shown in Table 8.2, and Figure 8.15 shows underwater discharging by the bubble knife in a protein solution and the microbubble line. The input power in this experiment was 1 W and discharging time was 20 ms. Figure 8.16 shows the experimental result of protein crystallization. Within 24 hrs., a large number of protein crystals were produced by the five-cycle discharging of the bubble knife compared to the control condition (without microbubbles).

Table 8.2 Experiment condition of reservoir solution and protein solution of each well.

	Compositions	Volume (μl)
Reservoir solution	5M NaCl	50
	1M Na acetate buffer pH4.5	5
	80% glycerol	31.2
	Pure water	13.8
Protein solution	Lysozyme(100mg/ml)	1
	Reservoir solution	2

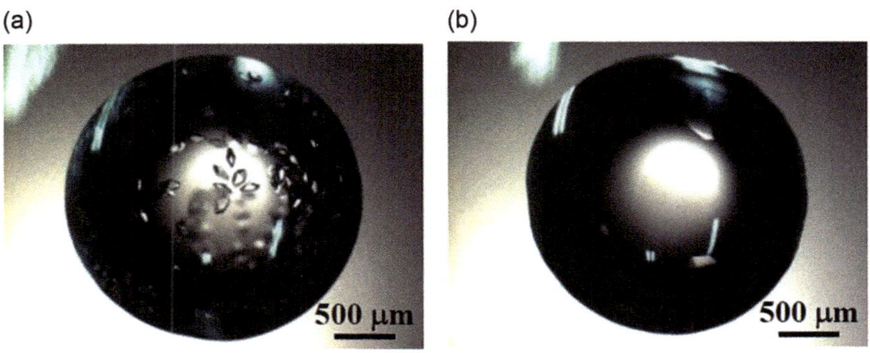

Figure 8.16 Comparison of the protein crystallization after 24 hrs. (a) Five cycles discharging of bubble knife. (b) Control. Marvelous number of protein crystal was induced by the bubble knife.

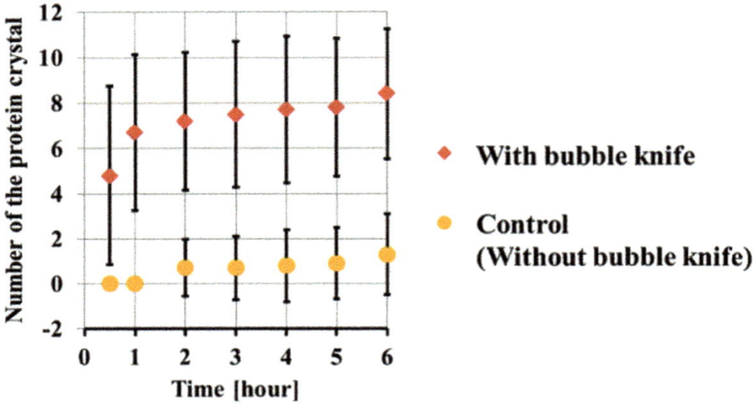

Figure 8.17 Comparison of the average number of the protein crystals (n = 10). Microbubble of bubble knife induced protein crystallization and enhanced growth speed.

Figure 8.17 shows time course to form protein crystal in each well. The protein crystals were observed within one hour in the solution discharged by bubble knife, while in the control condition, a protein crystal was not observed. After one hour increasing the rate, the number of protein crystal was nearly identical between the experimental result and control condition. Therefore, protein nucleation by electrically driven mono-dispersed microbubble occurred within one hour after discharging. From this result, we confirmed that the electrically driven bubble knife promoted protein crystallization. Figure 8.18 shows protein crystallization by bubble knife. Each condition was different in output power. Protein solution and reservoir solution condition are shown in Table 8.3. From

Table 8.3 Experiment condition of reservoir solution and protein solution of each well.

	Compositions	Volume (μl)
Reservoir solution	5M NaCl	50
	1M Na acetate buffer pH5.5	5
	80% glycerol	31.2
	Pure water	13.8
Protein solution	Lysozyme (100 mg/ml)	1
	Reservoir solution	1

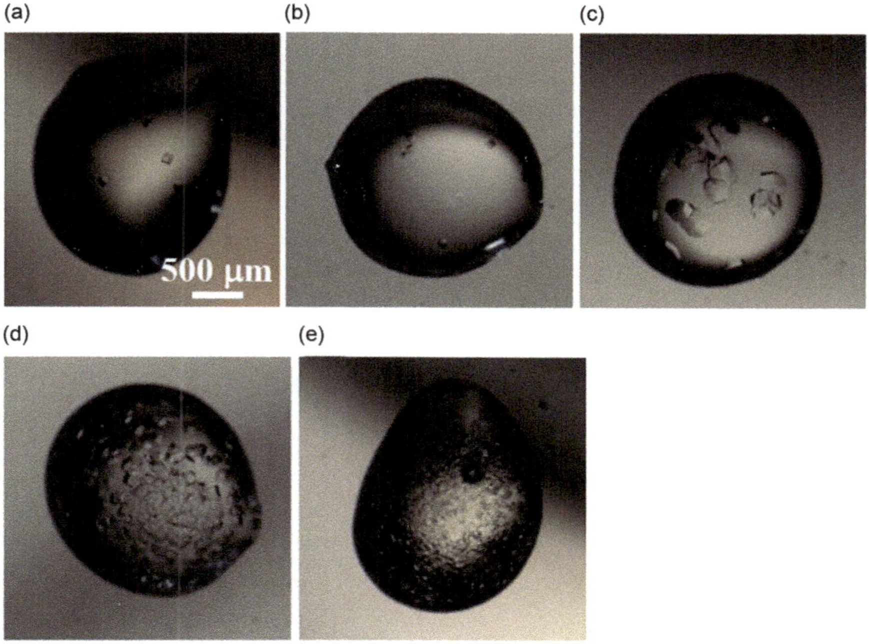

Figure 8.18 Comparison of the protein crystallization by 10 cycles discharging of bubble knife. Each condition was different with input power of bubble knife. (a) Control; (b) 0.8 W; (c) 1.0 W; (d) 1.5 W; (e) 2.0 W.

Figure 8.18, protein crystallization seemed to be promoted in the higher output power condition. It was observed that a number of smaller crystals appeared with the increase of applied voltage within a short period, while a limited number of crystals appeared when the applied voltage was small. Relatively, a large crystal growth condition was observed when the applied voltage was 1.0 V. This seemed to be because the ability of crystal core generation and low concentration of protein is balanced.

8.6 Protein Crystallization by Plasma-induced Bubbles

A recent report of plasma-assisted biological macromolecular crystallization [31] shows its possibilities to contribute to X-ray diffraction analysis aimed to understand the structure of proteins. However, the most important parameters of compositions of plasma (electrons, ions, radicals, and photons) are still unknown for crystallization. For the present study, the radical species are separated from irradiated plasma and extracted to expose to the protein solution by using reactive interface between plasma, water, and reagent solution based on the laminar flow of a microfluidic chip. The concept of the separation and extraction of radical species and its reaction to protein molecules is shown in Figure 8.19(a). Generally, plasma irradiation phenomenon yields several chemical or physical properties such as electron, ions, radicals, and photons. We focused on the radical species as important parameters to produce protein crystals, which may react with the protein molecules. To specify the parameters to accelerate protein crystallization, only plasma light was irradiated to the protein solution (Figure 8.19(b)) and the plasma jet [2] is also directly exposed to the protein solution (Figure 8.19(c)). It was confirmed that the

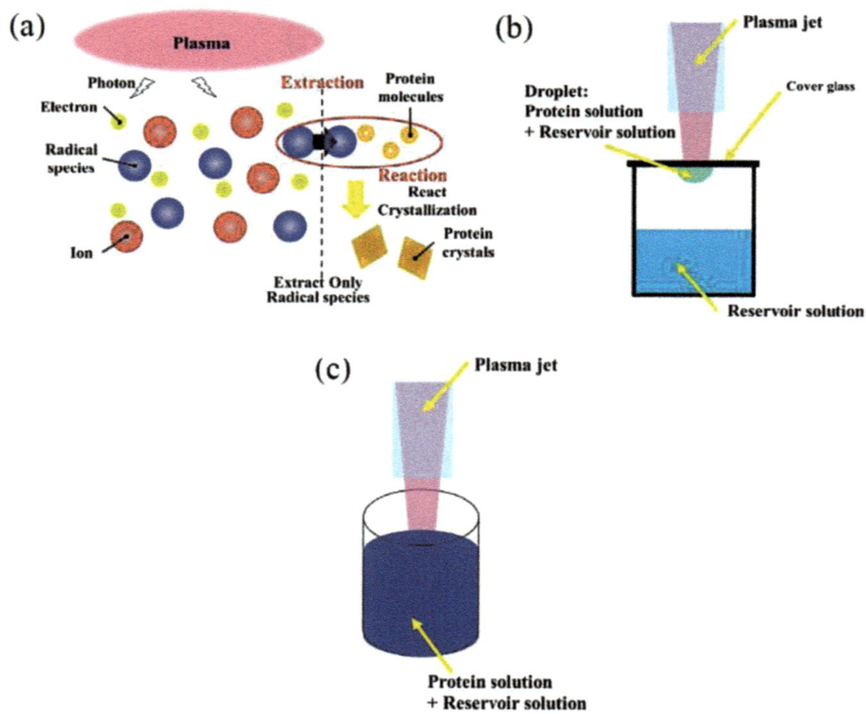

Figure 8.19 (a) Concept of protein crystallization by plasma-induced radical species; (b) only plasma light induced protein crystallization; (c) direct plasma jet exposure induced protein crystallization.

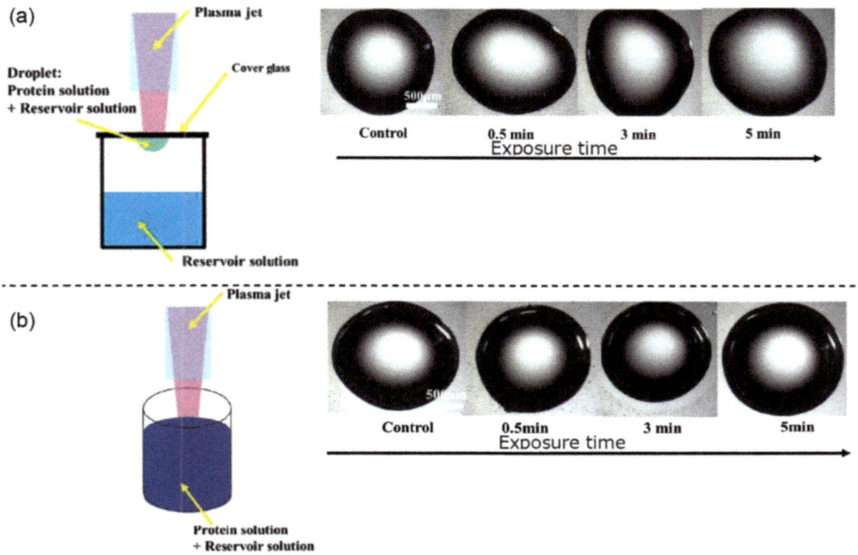

Figure 8.20 Protein crystallization by vapor diffusion method (hanging drop) as functions of exposure time. (a) Only plasma-light induced crystallization. (b) Direct plasma jet exposure induced protein crystallization.

both conditions (b) and (c) in Figure 8.19 did not produce any protein crystals, as shown in Figure 8.20(a) and (b). It seems that most of the plasma species such as electron, ions, and photons are not strong candidates to control the protein molecule. However, the lifetime of a radical is limited, so there is a possibility that the radical species could not actively react with the solution. Hence, two kinds of microfluidic chips are proposed to concentrate radical species, as shown in Figure 8.21. The radical concentration chips, which are "interface-reaction" chips and "circulation-bubble" chips, are proposed, as shown in Figure 8.21. Figure 8.22 shows the result of protein crystallization by using these chips. It was confirmed that both of the chips produced protein crystals successfully, and the circulation-bubble chip produce protein crystal more efficiently within a short time. This is because the radical-water interface of bubbles accelerate the crystallization. To evaluate the radical species, an ESR (electron spin resonance) measurement confirmed the existence of hydroxyl and other radicals, as shown in Figure 8.23. It is still unclear which radical species are most effective to the reagent; this is currently under investigation.

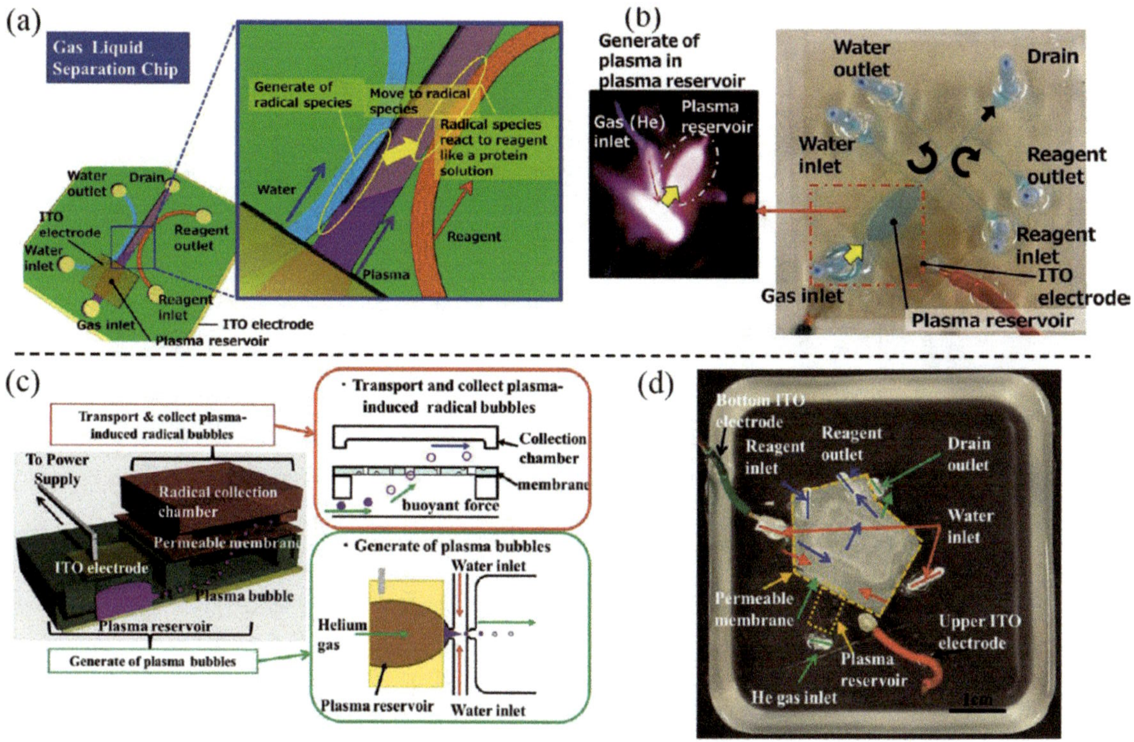

Figure 8.21 (a) Proposed microfluidic chip for transportation plasma-induced radical species by interfaces of water, plasma (radical), and reagent of laminar flow (interface-reaction chip). (b) Fabricated interface reaction microfluidic chip. (c) Proposed microfluidic chip that concentrates the plasma-induced radical species by circulating flow through the permeable membrane (circulation-bubble chip). (d) Fabricated circulation bubble microfluidic chip.

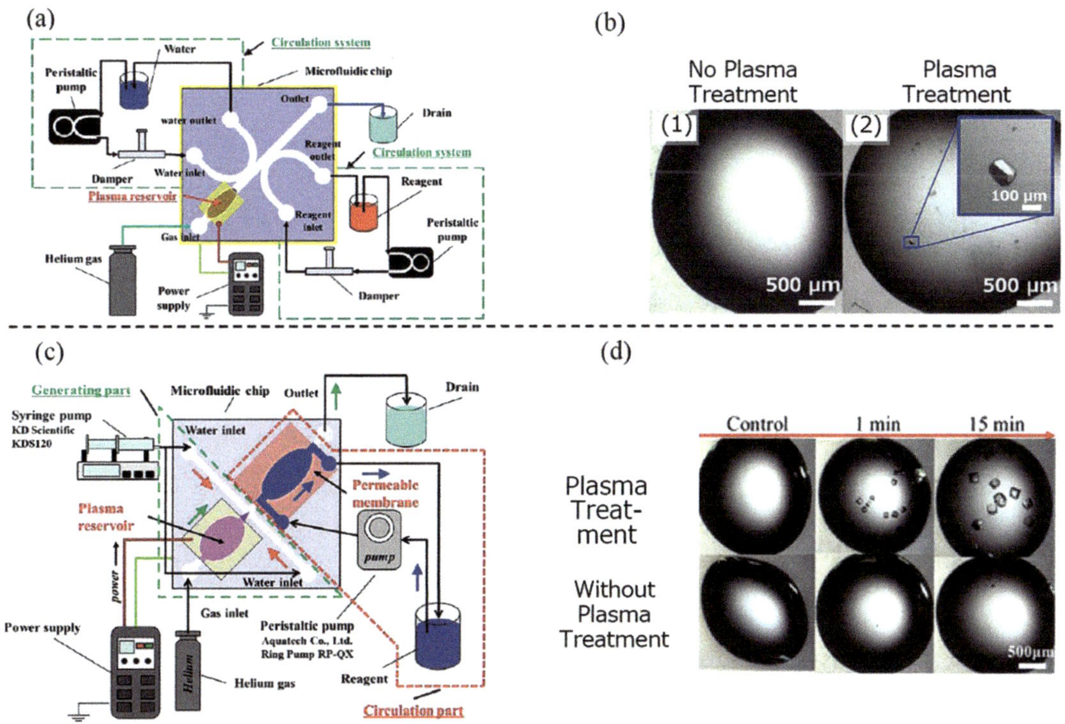

Figure 8.22 Production of protein crystals under the conditions: without plasma treatment and plasma-radical treatment: (a) experimental setup for interface-reaction chip, (b) protein crystallization by interface-reaction chip, (c) experimental setup for circulation-bubble chip, and (d) protein crystallization by circulation-bubble chip.

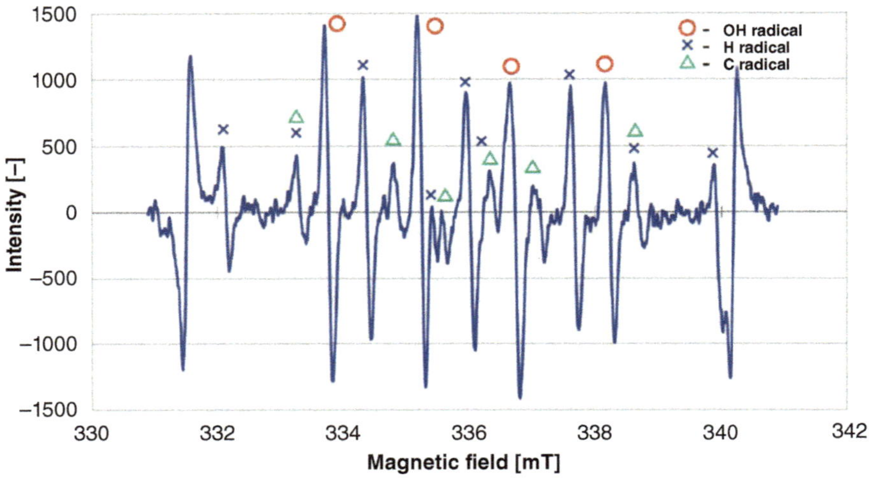

Figure 8.23 Identifying the hydrogen peroxide, C, and H radical in the circulation solution by the ESR.

Acknowledgments

This work was partly financed by the ministry of education, culture, sports, science, and technology (15H00903) and Japan Science and Technology Agency (JST) PRESTO program. The author thanks Prof. G. Kurisu at Osaka University, Prof. Umehara and Dr. Kikuchi at Riken, and Dr. Fukuyama for valuable advice on protein crystallization and Dr. Nakai at JEOL Ltd. for valuable advice related to ESR.

References

1 Makuta, T., Takemura, F., Hihara, E., Matsumoto, Y., and Shoji, M. (2006). Generation of micro gas bubbles of uniform diameter in an ultrasonic field. *J Fluid Mech* 2nd ed., 548: 113–131.

2 Garstecki, P., Fuerstman, M.J., Stone, H.A., and Whitesides, G.M. (2006). Formation of droplets and bubbles in a microfluidic T-junction –scaling and mechanism of break-up. *Lab on a Chip* 6: 437–446.

3 Barrero, A. and Loscertales, I.G. (2007). Micro- and nanoparticles via capillary flows. *Annu Rev Fluid Mech* 39: 89–106.

4 Yamanishi, Y., Feng, L., and Arai, F. (2010). On-demand production of emulsion droplets over a wide range of sizes. *Adv Robot* 24: 2005–2018.

5 Yokota, S., Yajima, F., Takemura, K., and Edamura, K. (2010). Electro-conjugate fluid jet-driven micro artificial antagonistic muscle actuators and their integration. *Adv Robot* 24: 1929–1943.

6 Ikeuchi, M., Tane, R., and Ikuta, K. (2012). Electrospray deposition and direct patterning of polylactic acid nanofibrous microcapsules for tissue engineering. *Biomed Microdevices* 14: 35–43.

7 Yeo, L.Y., Gagnon, Z., and Chang, H.-C. (2005). AC electrospray biomaterials synthesis. *Biomaterials* 26: 6122–6128.

8 Hagiwara, M., Kawahara, T., Yamanishi, Y., and Arai, F. (2011). Precise control of magnetically driven microtools for enucleation of oocytes in a microfluidic chip. *Adv Robot* 25 (8): 991–1005.

9 Inomata, N., Mizunuma, T., Yamanishi, Y., and Arai, F. (2011). Omnidirectional actuation of magnetically driven microtool for cutting of oocyte in a chip. *J Microelectromech Syst* 20: 383–388.

10 Zeng, F., Rohde, C.B., and Yanik, M.F. (2008). Sub-cellular precision on-chip small-animal immobilization, multi-photon imaging and femtosecond-laser manipulation. *Lab on a Chip* 8: 653–656.

11 Teramoto, J., Yamanishi, Y., Magdy, E.S., Hasegawa, A., Kori, A., Nakajima, M., Arai, F., Fukada, T., and Ishihama, A. (2010). Single-bacterial cell assay of promoter activity and regulation. *Genes to Cells* 15: 1111–1122.

12 Tseng, A.A., Chen, K., Chen, C.D., and Ma, K.J. (2003). Electron bean electron beam lithography in nanoscale fabrication: Recent development. *IEEE Trans Electro Packag Manuf* 26 (2): 141–149.

13 Hosokawa, H., Shimojima, K., Chino, Y., Yamada, Y., Wen, C.E., and Mabuchi, M. (2003). Fabrication of nanoscale Ti honeycombs by focused ion bean. *Mater Sci Eng* A344: 365–367.

14 Palanker, D., Nomoto, H., Huie, P., Vankov, A., and Chang, D. (2010). Anterior capsulotomy with a pulsed-electron avalanche knife. *J Cataract Refract Surg* 38: 128–132.

15 Palanker, D., Vankov, A., and Jayaraman, P. (2008). On mechanisms of interaction in electrosurgery. *New J Phys* 10: 1–15.

16 Loh, S., Carlson, G., Chang, E., Huang, E., Palanker, D., and Gurtner, G. (2009). Comparative healing of surgical incisions created by the peak plasmablade, conventional electrosugery, and a scalpel. *Plast Reconstr Surg* 124 (6): 1849–1859.

17 Palanker, D., Miller, J., Marmor, M., Sanislo, S., Huie, P., and Blumenkranz, M. (2001). Pulsed electron abalanche knife (PEAK) for intraocular surgery. *Investig Ophthalmol Vis Sci* 42 (11): 2673–2678.

18 Obreschkow, D., Tinguely, M., Dorsaz, N., Kobel, P., de Bosset, A., and Farhat, M. (2011). Universal scaling law for jets of collapsing bubbles. *Phys Rev Lett* 107: 204501.

19 Yamanishi, Y., Kuriki, H., Sakuma, S., Onda, K., and Arai, F. (2012). Local ablation by micro-electric knife. *IEEE Nanotechnol Mag* 6: 20–24.

20 Rayleigh, J.W.S. (1945). *The theory of sound*. Dover Publications, p. 344.

21 Walmsley, A.D., Laird, W.R.E., and Williams, A.R. (1985). Gas bubble fragmentation in an ultrasonic field. *Ultrasonics* 23: 170–172.

22 Chen, Y.-J. et al. (2003). Diagnosis of Oxide Films by Cavitation Micro-Jet Impact *Mater Trans* 44 (2): 327–335.

23 Takahashi, M. et al. (2005). Zeta potential of microbubbles in aqueous solutions: electrical properties of the gas-water interface *J Phys Chem B* 109: 21858–21864.

24 Takahashi, M. et al. (2007). Free-radical generation from collapsing microbubbles in the absence of a dynamic stimulus. *J Phys Chem B* 111: 1343–1347.

25 Congreve, M., Murray, C.W., and Blundell, T.L. (2005). Structural biology and drug discovery. *Drug Discov Today* 10: 895–907.

26 Chayen, N.E. and Saridakis, E. (2002). Protein crystallization for genomics: Towards highthroughput optimization techniques. *Acta Cryst* D58: 921.

27 Chayen, N.E. (1998). Comparative studies of protein crystallization by vapour-diffusion and microbatch techniques. *Acta Cryst* D54: 8.

28 Adachi, H., Takano, K., Hosokawa, Y., Inoue, T., Mori, Y., Matsumura, H., Yoshimura, M., Tsunaka, Y., Morikawa, M., Kanaya, S., Masuhara, H., Kai, Y., and Sasaki, T. (2003). Laser irradiated growth of protein crystal. *Jpn J Appl Phys* 42 (798).

29 Kam, Z., Shaikevitch, A., Shore, H.B., and Feher, G. (1980). Crystallization processes of biological macromolecules as a part of Electron Microscopy at Molecular Dimensions. *Proc book Life Sci Springer* 302–308.

30 Kuriki, H., Yamanishi, Y., Sakuma, S., Akagi, S., and Arai, F. (2013). Local ablation of a single cell using micro/ nano bubbles. *J Robot Mechatron* 25 (3): 476–483.

31 Ito, T. et al. (2011). Plasma-Assisted Biological Macromolecular Crystallization. *Appl Phys Exp* 4: 026201.

Index

Analytical Techniques for the Elucidation of Protein Function, First Edition. Edited by Isao Suetake, Rohit K. Sharma, and Hironobu Hojo.

t